Technoethics and the Evolving Knowledge Society:

Ethical Issues in Technological Design, Research, Development, and Innovation

Rocci Luppicini
University of Ottawa, Canada

T0338593

A volume in the Advances in
Information Security, Privacy,
and Ethics (AISPE) Book Series

Information Science
REFERENCE

Director of Editorial Content:	Kristin Klinger
Director of Book Publications:	Julia Mosemann
Development Editor:	Christine Bufton
Publishing Assistant:	Kurt Smith
Typesetter:	Deanna Zombro
Quality Control:	Jamie Snavely
Cover Design:	Lisa Tosheff

Published in the United States of America by
Information Science Reference (an imprint of IGI Global)
701 E. Chocolate Avenue
Hershey PA 17033
Tel: 717-533-8845
Fax: 717-533-8661
E-mail: cust@igi-global.com
Web site: http://www.igi-global.com

Library of Congress Cataloging-in-Publication Data

Luppicini, Rocci.
  Technoethics and the evolving knowledge society : ethical issues in technological design, research, development, and innovation / by Rocci Luppicini.
      p. cm.
  Includes bibliographical references and index.
  Summary:"This book introduces the reader to the key concepts and issues that comprise the emerging field of Technoethics, the interdisciplinary field concerned with all ethical aspects of technology within a society shaped by technology"--Provided by publisher.
  ISBN 978-1-60566-952-6 (hardcover) -- ISBN 978-1-60566-953-3 (ebook) 1.
Technology--Moral and ethical aspects. I. Title.
  BJ59.L87 2010
  174'.96--dc22
                          2009025606

This book is published in the IGI Global book series Advances in Information Security, Privacy, and Ethics (AISPE) Book Series (ISSN: 1948-9730; eISSN: 1948-9749)

British Cataloguing in Publication Data
A Cataloguing in Publication record for this book is available from the British Library.

All work contributed to this book is new, previously-unpublished material. The views expressed in this book are those of the authors, but not necessarily of the publisher.

Advances in Information Security, Privacy, and Ethics (AISPE) Book Series

ISSN: 1948-9730
EISSN: 1948-9749

MISSION

In the digital age, when everything from municipal power grids to individual mobile telephone locations is all available in electronic form, the implications and protection of this data has never been more important and controversial. As digital technologies become more pervasive in everyday life and the Internet is utilized in ever increasing ways by both private and public entities, the need for more research on securing, regulating, and understanding these areas is growing.

The **Advances in Information Security, Privacy, & Ethics (AISPE) Book Series** is the source for this research, as the series provides only the most cutting-edge research on how information is utilized in the digital age.

COVERAGE

- Access Control
- Device Fingerprinting
- Global Privacy Concerns
- Information Security Standards
- Network Security Services
- Privacy-Enhancing Technologies
- Risk Management
- Security Information Management
- Technoethics
- Tracking Cookies

IGI Global is currently accepting manuscripts for publication within this series. To submit a proposal for a volume in this series, please contact our Acquisition Editors at Acquisitions@igi-global.com or visit: http://www.igi-global.com/publish/.

Titles in this Series

For a list of additional titles in this series, please visit: www.igi-global.com

www.igi-global.com

701 E. Chocolate Ave., Hershey, PA 17033
Order online at www.igi-global.com or call 717-533-8845 x100
To place a standing order for titles released in this series,
contact: cust@igi-global.com
Mon-Fri 8:00 am - 5:00 pm (est) or fax 24 hours a day 717-533-8661

Table of Contents

Section 1:
Introduction to the Knowledge Society and Technoethics

Chapter 1

Chapter 2

Section 2:
Technoethics Deployment Framework in the Knowledge Society

Section 3:
Technoethics and Its Deployment

Section 4:
Globalization and Technoethics

Section 5:
Future Trends: Shaping the Future of Technoethics

Preface

INTRODUCTION

The purpose of *Technoethics in an Evolving Knowledge Society* is to introduce the reader to the key concepts and issues that comprise the emerging field of Technoethics. Following from the *Handbook of Research on Technoethics*, the term "Technoethics" is used in this book to define the interdisciplinary field concerned with all ethical aspects of technology within a society shaped by technology. It deals with human processes and practices connected to technology which are becoming embedded within social, political, and moral spheres of life. It also examines social policies and interventions occurring in response to issues generated by technology development and use. This includes critical debates on the responsible use of technology for advancing human interests in society. To this end, it attempts to provide conceptual grounding to clarify the role of technology in relation to those affected by it and to help guide ethical problem-solving and decision making in areas of activity that rely on technology (Luppicini, 2008).

Regarding the scope of the field, Technoethics, denotes a broad range of ethical issues revolving around technology—from specific areas of focus affecting professionals working with technology to broader social, ethical, and legal issues concerning the role of technology in society and everyday life. Specific areas of focus connected to Technoethics form the following branches: Computer and Engineering Ethics, Educational Technoethics, Information and Communication Technoethics, Technoethics and Society, Technoethics and Cognition, and Biotech Ethics. Broader issues concerning social, ethical, and legal implications role of technology are addressed under topic areas such as intellectual property, legal rights and responsibilities, codes of conduct, digital divide, access, privacy issues, ethical dilemmas, free speech, criminal activities, security, regulations, policy planning, equity, risk management, etc. Because of the interdisciplinary nature of Technoethics, this book attempts to integrate the core topics of interest concerning ethics and technology into a comprehensive whole.

Objectives

This book is about Technoethics. The overall goal of the book is to provide theoretical and practical explanations of how to use Technoethics in technology design, development, research, innovation, and application. It focuses on technoethical inquiry within multiple disciplines and fields by To this end, the objectives are threefold: (1) provide readers with a comprehensive overview of key developments and guiding principles driving the field of Technoethics to inform technoethical inquiry, (2) to delve into core niche areas of Technoethics derived from multiple fields and disciplines where social and ethical aspects of technology are of concern, and (3) to advance theoretical perspectives and model to help guide technoethical decision-making within a technologically advances knowledge society. Given the rapid growth of technology in society, this book provides an invaluable set of tools to deal with a broad range of 'new' ethical concerns possible within a knowledge society defined by technological progress and human innovation.

Importance

The transforming power of science and technology within the evolving knowledge society is linked to growing ethical problems Moor's Law asserts that ethical problems in society increase as the influence technology increases (Moor, 2005, pg. 117). Moor argued that the rise of powerful technologies provides new opportunities for action which require the development of well thought out ethical policies. In this sense, the study of technology has a double importance because it is both a major catalyst in the evolving knowledge society and an intricate part of societal development affected by social and ethical values. This places the field of Technoethics and the various forms of technoethical inquiry at the nexus of progress within the evolving knowledge society.

Because technology has such a massive influence on human life, it is crucial to understand the relevant ethical questions in early stages in scientific and technological developments. This book serves a vital function in helping to offer explanations and engage the reader in critical debates connected to the ethical dimensions of technology in society. It helps readers from multiple fields and disciplines engage with ethical dilemmas posing serious challenges for researchers and practitioners affected by technology.

This is book dedicated to the study of technology and ethics. There is a growing demand for such a text to help guide technology designers, information specialists, researchers, engineers, managers, and administrators in areas where social concern over technology is highlighted. Current research and theory in Technoethics is useful for assessing the ethical use of technology, guarding against the misuse of technology, and formulating common principles to guide new advances in techno-logical development and application to benefit society. To this end, Technoethics is important and indispensable aspect of technological development that helps to clarify its role in meeting the demands of those affected by it.

In terms of intended audience, this book is written for undergraduate and gradu-ate level students in social and technical studies, along with researchers and other professionals working in the field of Technoethics. The combination of pertinent theoretical/philosophical perspectives, comprehensive reviews of core literature, and practical tools (i.e., technology assessment and design strategies) discussed in this book provide a balanced reading on contemporary work on ethics of technol-ogy for technical students and students studying ethics and technology within the social sciences. The addition of selected case studies at the end of this book will be of particular interest to students and researchers interested in conducting research in the field of Technoethics. These case oriented chapters offer the reader relevant studies of selected topic areas on ethics and technology which go beyond the lit-erature review aspect and offer some direction on technoethical inquiry to guide current and future work. Moreover, the appendices include several interviews from leading experts working in controversial areas of technology and ethics. This will be of special interest to technology experts and other professionals working closely with new technologies. It is hoped that all readers will gain a thorough understand-ing of the scope and importance of ethics and technology in society which drives the field of Technoethics.

ORGANIZATION OF THE BOOK

This book is organized into five parts with a total of fourteen chapters. The five parts help to organize the book into logical areas of focus. Section 1, Introduction to the Knowledge Society and Technoethics, introduces the reader to the concep-tualization of the knowledge society with an emphasis on technology and ethical considerations entrenched within human activity and culture at its core. Section 2, Technoethics Deployment Framework in Knowledge Society, provides key theoreti-cal and historical grounding for Technoethics. It traces out the multiple knowledge bases which helped shape Technoethics today. Section 3, Technoethics and Its De-ployment, provides chapters dealing with Technoethics in core areas of work, life, and society affected by new technology. Section 4, Globalization and Technoethics, provides sample studies in selected areas to help demonstrate how Technoethics is currently being applied . Finally, Section 5, Future Trends: Shaping The Future of Technoethics, provides a look ahead at selected challenges and opportunities for advancing Technoethics and the ethical use of technology in society.

The chapters are arranged within each part of the book to provide readers with essential background information before engaging in specific issues in Technoethics. Given the converging trends in many areas of technology (i.e., computing, information sciences, communications) some repetition that flows from chapter to chapter is unavoidable. However, the book attempts to minimize repetition and focus on areas of Technoethics that are the most relevant to students in social and technical studies. The specific focus of each chapter is as follows:

- Chapter 1 provides an overview of the knowledge society with an emphasis on the role of technology in today's knowledge society marked by powerful technologies that reshape the economy, culture, politics, laws, and ways of living. It provides a review of key developments driving the knowledge society while attempting to demonstrate how technology is interwoven within the context of societal changes that highlight the centrality of technology and knowledge building in work and life.

- Chapter 2 connects the field of Technoethics with its philosophical and historical context. It sketches out the rise of Applied Ethics which brought philosophical inquiry into the realm of real world human practices and the pursuit of science and technology. It then reviews pivotal work from the Philosophy of Technology and technocritical writings to demonstrate how scholars were became more aware of the need to connect technological development to human social and ethical entrenched in how people live, work, and play. This placed ethical questions related to technology on new ground requiring special status, which led to the emergence of Technoethics.

- Chapter 3 explains technological consciousness and moral agency (human and artificial). To this end, it discusses the conceptualization of technology that has become narrowed through recent conceptualizations of the term. In doing so, it presents a more comprehensive view of technology deeply entrenched in human history, society, and mind. In doing so, it attempts to expel common myths about technology and consciousness to provide more suitable base for scholarly inquiry and debate in technology studies and its ethical aspects. In addition, it discusses moral agency and the possibility of artificial moral agency.

- Chapter 4 frames technoethical inquiry as a social systems theory and methodology. It explains how systems methodology provides powerful tools to help integrate interdisciplinary expertise (and experts) and deal with competing values and ethical issues connected to the actual and potential influences of technology in society. The chapter applies principles of technoethical inquiry to applications in technology assessment and technology design in an effort to highlight the practical's implications of adapting Technoethics.

- Chapter 5 discusses the nature of technological systems in society and how the success of technoethical inquiry depends, in part, on appreciating the multifaceted nature created of a technological system at different levels of technological system operations. It does so by providing a comprehensive overview of key concepts and guiding principles driving the field of Technoethics to inform technoethical inquiry. The chapter also provided a breakdown of core branches of Technoethics, namely, Computer and Engineering Technoethics, Technoethics and cognition, Information and Communication Technoethics, Educational Technoethics, Biotech Ethics, and Technoethics and Society.

- Chapter 6 deals with the emergence of Information and Communication Technoethics as an interdisciplinary research that focuses on ethical and social aspects of advancing information and communication technologies in all areas of life and society. It begins with a background sketch of Information and Communication Technoethics rooted in technological developments in contemporary society, along with key academic developments within information ethics, technology studies, interdisciplinary studies, and philosophy. Then, it reviews current work on ethical aspects of information and communication technology that increasingly mediates important aspects of how individuals live and interact.

- Chapter 7 explores Biotech Ethics and reviews key technoethical speciality areas guiding the ethical use of technology within the context of modern healthcare and medicine. By examining key developments, public concerns, and the role of ethics committees, this chapter provides information on key areas of concern and possible applications. By drawing on a technoethical framework, this chapter makes suggestions to help protect the public while guiding new developments in biotechnology that ensure mutual benefit-sharing.

- Chapter 8 discusses key developments in Engineering and Computer Ethics. It focuses on the establishment of various engineering professional bodies to ensure that engineers were responsible for potentially harmful constructions and the subsequent creation of codes of engineering ethics to help guide professional conduct. It explores major industrial developments in the 20th century and the rise of ethical implications and demand for improved codes of ethics in response to controversial technological practices (e.g., human caused environmental disasters, military weapons, nanotechnology advances, etc.)

- Chapter 9 dealswith Educational Technoethics as an important branch of technoethics dedicated to social and ethical aspects of technology within traditional educational settings as well as professional educational contexts. The chapter discusses general educational aspects (concerned with ethical use of technology to promote the aims of education) and professional aspects (a specialized area focusing on the development and evaluation of ethical codes and standards) to guide decision-making and training about technology in education and professional life.

- Chapter 10 introduced Technoethics and Society as an important area of inquiry focused on the ethical use of technology to promote the aims of individuals, organizations, and global development trends. It provides a spotlight on technology within a global context of human activity shaped by a variety of interconnected influences. In articulating global technoethics, this chapter places technology and human activity at the core of globalization.

- Chapter 11 is a case oriented chapter which presents a research study exploring the abuse of electronic technologies in the workplace and use of counter measure electronic technologies within organizations. Findings highlighted how many students entering the workforce openly admitted to electronic technology abuse at work and did not believe there to be an ethical problem unless it interferes with work productivity. This view was supported by research scholarship on increasing technological convergence within contemporary culture and the presence of social values which link power and prestige to technology connectedness and constant information access. Findings suggest that an ethical dilemma created by the conflict of ethical beliefs and social norms with organization rules and regulations prohibiting personal electronic technology use in the workplace.

- Chapter 12 is another case oriented chapter which presents a case study of ICT development in South Africa. The case explores how technoethical considerations are intertwined with historical, political, and social factors that have impeded progress in ICT development initiatives. The chapter argues that globalization efforts in areas where technology and ethical considerations are key must invest greater attention to developing multidimensional theoretical frameworks such as technoethical models to help explain key drivers, processes, players, and consequences of globalization.

- Chapter 13 is a case oriented chapter which focuses on the cultural tensions that arise when a technology rich culture threaten the sustainability of a technology poor culture. A technoethical study of cultural tensions between aboriginal people and dominent French and English Canadian populations. This study explores how technoethical considerations are intertwined with historical, political, and social factors that have threatened the sustainability of aboriginal culture in Canada. Findings suggest that more attention must be invested to ensure that that globalization efforts by technology rich dominant cultures do not lead to the demise of technology poor marginalized cultures.

- Chapter 14 culminates the book with an epilogue discussing the new world of technoethics and the challenges of creating and managing scientific and technological knowledge for an evolving knowledge society

Given the rapid growth of technology book provided a glimpse at the latest developments in Technoethics to deal with a broad range of 'new' technology advances possible The fourteen chapters delve into the broad array of technoethical concerns faced by individuals within an evolving knowledge society. As an added resource, several invited experts in Technoethics from around the globe were interviewed to provide insider information on key challenges and opportunities in a variety of areas where new technologies are debated. These interviews, which can be found in the appendices of this book, are intended to provide useful insight into what works and what does not work when dealing with controversial technology.

Acknowledgment

Without the continual support of IGI Global, *Technoethics in an Evolving Knowledge Society* would not have been possible. This author would also like to acknowledge the important contributions from Mario Bunge, a pioneer in Technoethics. This book was inspired from the work undertaken by over 100 contributors to the *Handbook of Research on Technoethics* (2008) and ongoing contributions from the editorial and review board of the *International Journal of Technoethics* (forthcoming in 2010). A special thanks go to all those who provided constructive feedback and critical comments on earlier versions of this book. Nevertheless, this author takes full responsibility for any errors, omissions, or weaknesses in this work. It is hoped that this work provides solid grounding for developing future scholarship as the field continues to evolve.

Sincerely,
Rocci Luppicini, University of Ottawa

Section 1
Introduction to the Knowledge Society and Technoethics

Chapter 1
The Knowledge Society

INTRODUCTION

Gone are the days of the divine rights of kings—or of anyone else, whether owner, manager, labor leader, politician, bureaucrat, technologist or scholar. Absolute and groundless authority is being contested all over the world: ours is an iconoclastic age --Bunge, 1977, p. 96.

Generally speaking, all societies in history were knowledge societies. However, the modern, conceptualization of the "knowledge society" can be traced to John Stuart Mill's (1831) *The Spirit of the Age* where social progress was explained through the diffusion of knowledge (intellectual wisdom) and increased opportunities for

DOI: 10.4018/978-1-60566-952-2.ch001

individual choice arising from industrialization. This was an early indicator fore-shadowing the transformation of modern society into a knowledge society. Beginning in the early 20th century, industrialized nations became increasingly reliant on economic investment in the production and distribution of knowledge in training, education, work, research and development (Abramovitz & David, 2000). Also, the importance of knowledge in society became even more pronounced through the advent of specialized areas of science and technology in society. As stated by Stehr (2002), "Contemporary society may be described as a knowledge society based on the extensive penetration of all its spheres of life and institutions by scientific and technological knowledge."

New scientific and technological developments are transforming society into a knowledge society as they become deeply embedded in popular culture, private and public affairs, work and educational settings, social practices, and public institutions. Knowledge derived from scientific and technological developments is redefining key aspects of social life including: How scientific discoveries are treated, how health care is distributed, how children learn, how adults interact and work with one another, how governments conducts affairs, how ethnic groups preserve their culture, how medical research is controlled, how business transactions are conducted, and how nations interact within a global marketplace. Moreover, the growing body of knowledge derived from scientific and technological developments has the potential to provide important social benefits such as reversing environmental damage caused in the past, reducing pollution in the present, and preserving natural resources for the future. Due to the propagation of powerful new scientific and technical advances within this knowledge society, there is a need for a study of social and ethical aspects of such advances to leverage benefits and guard against the misuse of new tools and knowledge. This is not an easy task and requires a clear understanding of key terms and concepts.

The notion of "knowledge society" is both powerful and complex. It is powerful because it has received widespread attention as both a chief development and challenge for public and private institutions, leaders in industry, and governments with vast amounts of resources and control in decisions leading to large-scale innovations directing societal growth. This is particularly salient in the domain of economics where terms like "knowledge-based economy" and "knowledge management" represent core areas of theory and research. The notion of 'knowledge society' is complex because has many facets and supporting constructs connected to a multidisciplinary based of scholarship. It is the focus of scholarly attention in multiple disciplines and fields of study from economics to sociology to information sciences to business. Other terms such as "information society", "postindustrial society" and "posthuman society" are related terms that help draw attention to the major shift affecting developed and developing nations around the world as we move into the 21st century.

This chapter traces the concept of the "knowledge society" from its recent roots in economics to subsequent developments which grew out of other disciplines and fields including, business, organizational management, science and technology studies, and information science. More specifically, the goal of this chapter is to review the coming of the knowledge society. It is also to expose the ethical dimensions underlying mechanisms by which knowledge is created, managed and applied in society.

BACKGROUND

Historical Background

The concepts of "knowledge "and "knowledge society" are far from new. Since before the time of Socrates, knowledge has been important as both a subject of scholarly pursuit and a core component of economic development in society. The advancement of knowledge to create new products, services, and innovations has been an important component of societal growth from metal forging in early Greak civilization to corporate control of media outlets in the 21st century. However, the placement of knowledge at the center of economic and societal growth is a relatively recent phenomenon marked by a shift in the modern world from an industrial age to an information age. In developed countries like the United States, this shift occurred in the 1960s and 1970s with the rise of the knowledge-based economy.

Fritz Machlup's (1962)*The Production and Distribution of Knowledge in the United States* was a pioneering work which placed knowledge at the forefront of economic focus. Machlup (1962) provided the first comprehensive analysis of the U.S. economy that separated out the unique contribution of knowledge production as part of the overall gross national product. Under this framework, knowledge production was statistically calculated from key indicators(education, research and development, communication and media, information technology, and information services) which constituted overall knowledge production (knowledge industry). This pioneering work paved the way for subsequent analyses placing knowledge production at the center of society from an economic standpoint as is seen in Drucker (1968).

Peter Drucker was another driving force behind the knowledge society and coined the term "knowledge worker" to describe jobs in society focusing largely on the development or use knowledge in the workplace (Drucker, 1959). He also defined the idea of the knowledge-based economy (the use of knowledge to produce economic benefits) which helped further solidify interest in the economic aspects of knowledge as a core component of the evolving knowledge society. Peter Drucker's (1968)*The Age of Discontinuity* built on Machlup's study and predicted that that knowledge production would account for over 50% of overall gross national production by the 1970s. This early work from scholars like Machlup and Drucker nurtured in the notion of the knowledge society in economics.

By 2006, the knowledge economy was estimated to represent 75% of total work within North America (Haag, Cummings, McCubbrey, Pinsonneault, & Donovan, 2006). Kahin and Foray's (2006)*Advancing Knowledge and the Knowledge Economy* is a particularly useful collection of essays that examine how processes for creating and organizing knowledge intersect with technology to transform social and economic conditions within the knowledge society. This intertwining of knowledge and technological innovation. is particularly salient with the continued growth of "ambient" and "ubiquitous" technologies designed to be seamlessly integrated into areas of work and life (Aarts & Encarnação, 2006). Increasingly intelligent and intuitive interfaces are being built into everyday objects (smart homes, smart networked objects, and wearable computing) which can respond to individual behaviors in sophisticated ways (Aarts & Encarnação, 2006).

The focus on knowledge in society spearheaded in economics spread to other disciplines and fields under a variety of terms such as post-industrial society (Bell, 1976), postmodern society (Lyotard, 1984), posthuman society (Fukuyama, 2007), information society (Garnham, 2004),network society (Barney, 2003), and information age (Castells, 2000). On the one hand, this growth helped bring attention to the many new considerations revolving around the centrality of knowledge in studying human activity within many areas of work and life. On the under hand, this spread of related terms has created confusion about the meaning of "knowledge society" and its unique contribution to the study of social and ethical aspects of society (i.e., technoethical inquiry).

Conceptualizing the Knowledge Economy

At term "knowledge society" is an important concept and one that must be clearly defined as to not confuse it with similar concepts with overlapping aspects. Closely related concepts to the knowledge society include, post-industrial society (Bell,1973), information society, and network society (Castells, 2000). In Daniel Bell's, (1973)*The Coming of the Post-Industrial Society*, the idea of a postindustrial society was used to highlight the economic shift in society from the production of goods to the production of services connected to technology designers and users. The notion of a post-industrial society came to denote new relations of science to technology within contemporary society. In a similar vein, Manuel Castells' (2000)*The Rise of the Network Society. The Information Age: Economy, Society and Culture* used the notion of information society and network society to highlight the impact of sophisticated network technology at a global level. He envisioned a global city as a distributed phenomenon transcending individual cities.

As will become evident in this book, the concept of 'knowledge society' is the most suitable term available for describing technoethical inquiry in society. First, although the terms are often used interchangeably, knowledge is more closely aligned with organized aspects of human life and society where science and technology are influential, such as knowledge management and knowledge economy. It more closely linked to core human activities within organizations and society (knowledge management, knowledge organization, knowledge creation, knowledge economy) whereas information/network society are more focused on ICT, which is one aspect of technology and technique which does not address environmental or economic considerations as clearly as knowledge does.

Second, given the importance attributed to tacit and explicit knowledge building activity as core to organizational development, the knowledge society is a better fit to describe organized forms of scientific and technological activity over competing terms. The label of knowledge society is versatile because, tacit and explicit dimensions of knowledge have been mapped onto to human activity at an individual and organizational levels. Third, in addition to the close alignment with core technological activity shaping society, and practical versatility regarding individual and organizational knowledge, a knowledge society represents the most suitable framework for framing the plethora of ethical issues connected to technology and society. The notion of a competing term, network society, fails as a metaphor to accommodate technoethical concerns outside the scope of developed nations where networked technologies are widespread. Fourth, despite the unique appeal of various terms employed in the literature, knowledge remains a central focus and defining feature of economy and society. Lyotard (1984) asserted that, "knowledge has become the principle force of production over the last few decades". This is partly due to the entrenchment of knowledge as a valuable commodity in the economy and areas of human activity. "In a Third Wave economy, the central resource – a single word broadly encompassing data, information, images, symbols, culture, ideology, and values – is actionable knowledge" (Dyson, Gilder, Keyworth, & Toffler, 1994). This idea follows Hardt and Negri's (2000)*Empire* in linking the production of knowledge and cultural artifacts to the core work of individuals involved in a growing immaterial labour force in society. As stated by Nico Stehr (2002) "Contemporary society may be described as a knowledge society based on the extensive penetration of all its spheres of life and institutions by scientific and technological knowledge" (18). For these reasons, knowledge society is the preferred context of technoethical inquiry, defined as a society where the creation, management, and application of knowledge is central to social, economic, political, and cultural activity.

KNOWLEDGE SOCIETY AND THE CULTURE OF TECHNOLOGY

One of the biggest challenges within the knowledge society is understanding culture and how it sustains itself (or fails to) amidst rapid technological growth, globalization, and the reorganization of work and life. Part of the challenge lies in appreciating how technological culture has a broad reaching impact with a plethora of consequences within the knowledge society that influences human life, organizations, and global development. A small but growing body of scholarship developed over the last fifty years has helped highlight some of the ways in which technology has a transforming influence on human culture and society and visa versa (Castell 2000; Elzinga, 1998; Haraway, 1991; Jenkins, 2006; McLuhan, 1962,1964; Real, 1975; Turkle, 1995; Williams, 1975).

The importance of culture and technology was first recognized by Marshall McLuhan, who was responsible for popularizing the study of technology in Communications by drawing attention to the influence of modern communication technology on the human senses and understanding. McLuhan's (1962)*The Gutenberg Galaxy: The Making of Typographic Man*, attempts to show how communication technology influences the cognitive organization of sensory experiences, which in turn, alters the social world," [I]f a new technology extends one or more of our senses outside us into the social world, then new ratios among all of our senses will occur in that particular culture." McLuhan's (1964), *Understanding Media: The Extensions of Man* (1964) was a pioneering study of technology in the media which argued that the media, rather than the content, should be the main focus of study (the medium is the message). By drawing attention to the importance of studying the relation between media technology, the senses, and culture, McLuhan raised public awareness of the close connection and dependency humans have with technology and how this alters culture by changing how individuals interact with one another within society. This work also nurtured in a media ecology trend in the 1970s with a focus on technology mediation effects and the powerful role media technology plays in regulating meaning in life and society (Real, 1975, Williams, 1975).

In the 1980s and 1990s, a number of cultural and critical perspectives on technology in society have helped provide useful knowledge about underlying cultural meaning and power structures in society that often remain implicit and difficult to explain. Exemplar works in this area include Tichi's (1987)*Shifting Gears: Technology, Literature, Culture in Modernist America,* Turkle's (1995)*Life on the Screen:Identity in the Age of the Internet,* Haraway's (1991)*A Cyborg Manifesto:*

Science, Technology, and Socialist-Feminism in the Late Twentieth Century, and Franklin's (1990)*The Real World of Technology*. Tichi (1987) explores how American technology between the 1890s and 1920s influenced American culture through the language used by writers of the time (e.g., Ernest Hemingway and John Dos Passos). Ticcchi argued that technology was entrenched in the culture and the minds of poets and novelists in society. She noted, "There's nothing sentimental about a machine, and: A poem is a small (or large) machine made of words" (270). This work helped raise awareness about the entrenchment of technology has powerful transforming effects on culture and society that are implicit (see levels of technology section in chapter five)

The transforming role of technological culture is not limited to the outside world but extends into the virtual world of the Internet giving rise to new cultural forms (cyber-cultures). An important work on culture and society in the age of the Internet by Turkle (1995) explores how computer are forcing individuals to re-examine notions of identity in the age of the Internet. The text provides new insight into how widespread computer use has changed relationships, created new psychological meaning and even provided a new sense of identity in contemporary social. Franklin's (1990)*The Real World of Technology* also explored how technology changes social relationships by redefining conceptions of power and accountability. This work views technology as a form of practice (ways of doing something) embedded in society within various organizations, procedures, symbols, language acts and mindsets. In a slightly different vein, Haraway (1991) advanced a feminist approach to cyberculture, which advocated that technology liberates women from previously held notions of female embodiment and mothering. Haraway (1991) emphasizes, "Only by being out of place could we take intense pleasure in machines, and then with excuses that this was organic activity after all, appropriate to females". Other noteworthy analyses of technological culture focus on the power of information technology including Castell's (2000) The Information Age: Economy, Society and Culture Vol.I and Jenkins' (2006) Convergence Culture: Where Old and New Media Converge (see Chapter 6 for a more detailed discussion)

The interest in technological culture has also spread to the study of social institutions, organizations, and globalization. How does technology culture affect values, knowledge creation, and communications within social institutions and organization? What are the implications of a global technological culture? A number of studies have examined the institution of science and technology as a professional culture present within organizational work and professional activity (Knorr-Cetina, 1999; Latour & Woolgar, 1986). Latour and Woolgar (1986) explore the culture of scientific work in how scientists shape the co-construction of facts and meaning. Knorr-Cetina's (1999)*Epistemic Cultures: How the Sciences make Knowledge* offers an ethnographic exploration of how epistemic culture shape work practices, institutional organization, and power relations through the use of technologies. This type of work helps reveal how there is a culture (or cultures) of technology influencing contemporary organizations within society. At a macro level, globalization studies focus on technology and the changing landscape of cultural values, practices, images, languages, and people. Tomlinson's (1999)*Globalization and Culture and*Robertson's (1992)*Globalization: Social theory and Global Culture* explore the idea of a global culture in the face of technological advancement. This macro-level work is important in leveraging understanding about how global social processes spurred by technological developments circumnavigate geographical constraints on traditional cultural arrangements (see Chapter 10 and Chapter 11 for a more detailed discussion). Taken together, the cultural dimension of the knowledge society represents an important area of consideration for any social inquiry (including technoethical inquiry) that seeks to investigate human values and ethical considerations.

ETHICAL CHALLENGES

The rise of the knowledge society has also led to the identification of a number of challenges and ethical issues: Limitations of human knowledge processes, discrimination in the workplace, strained work and life balance in technologically enhanced work environments, inequalities in information access for parts of the population, unequal opportunities for scientific and technological development, organizational responsibility and accountability issues, reputation and trust building challenges, and intellectual property issues. First, although knowledge is a central part of the economy and human life, our limited capacity to remember, access, and utilize acquired knowledge restricts opportunities to reproduce it for advancing work and society. This is partly due to limited cognitive and organizational capabilities (Ashcroft, 1989; Polanyi, 1966).

Under information processing theory, working memory has a limited capacity and too much information can lead to cognitive overload resulting in loss of information from short term memory (Ashcraft, 1989). Polanyi's (1966) identification of tacit and explicit elements in knowledge demonstrate the challenge individuals have in making acquired knowledge explicit. Many experts rely on a high degree of procedural knowledge (knowledge how) in work but lack the inability to articulate it to themselves or others if needed, "I cannot explain it but I can show you." This limits the reproducibility of knowledge that lays locked up inside individual experts. Compounding this issue is the limited memory capacity of individuals who often struggle to articulate knowledge over long periods of time that was once accessible. Memories can fade and facts can become blurred over time. Middleton (2005) indicated, "We don't yet understand the role of tacit knowledge in the generation of new ideas other than to note that inventors report it consistently as a part of the invention process".

At the organizational level, the problem of accessing member knowledge for organizational can limit an organization's ability to innovate and respond to change. Beer's (1972)*Brain of the Firm* was a seminal text in organizational management that captured the complexity of organizational life and the challenges of knowledge transfer between individuals that make up the organization (brain). Not everything that goes on in the brain is at the level of consciousness just as not everything that goes on in organizations is accessible by organizational members that need it. These cognitive and organizational limitations become even more pronounced in the technological context where advanced ICT's accelerate knowledge production and further tax limited cognitive and organizational capacities central to knowledge reproduction. This can greatly limit the availability of knowledge needed to make ethical decisions concerning science and technology innovation.

Second, the knowledge society is intertwined with changing technology requiring new skills of its workforce. This can lead to expensive training needs for some workers and unfair hiring practices to limit training expenses, which may place parts of the population at an unfair disadvantage in adapting to a knowledge society. Public concern and debate over ageism in the workplace and other areas of society is a serious public concern that has spurred considerable debate over recent years. For instance, McCann and Giles (2002) argued that perceptions of resistance to technology are linked to prejudice and discrimination in the workplace with older workers. This is believed to be linked to commonly held stereotype that older workers have difficulty adapting to new technology (Breakwell & Fife-Schaw, 1988). There is the perception that that older workers lack experience with new technology and that retraining programs may be less effective and more expensive for older workers (Cutler, 2005).

Third, the advent of sophisticated information and communication technology (ICT) within a global economy has created more flexible work possibilities than ever before while leading to time and space compression within the knowledge society. In the knowledge society, physical distance loses its constraining power that in previous times limited where and when working could take place. The growth of virtual organizations (Cascio, 2000) with no real physical office space offer more flexible working conditions while traditional organizations are beginning to allow more opportunities for tele-work (working from home). The reason for this shift is attributed to multiple factors including, increased access to global markets, more efficient customer services, and reduced real estate expenses (Cascio, 2000). This blurring of the traditional time and space boundaries has also led in many cases the blurring of work and personal life (Saetre & Sornes, 2006) and a longstanding public debate concerning job elimination, deskilling workers, work overload, increased stress, and increased work related mental problems (Aronowitz & DiFazio 1994; Turnage, 1990). Working in the global economy can create a disembeddedness of organizations and people which challenges efforts to building meaning and a sense of organizational affinity with working life (Anderson, & Englehardt 2001).

Fourth, access to information and knowledge resources within a knowledge society tend to favour the economically privileged who have greater access to the technological tools needed to access information and knowledge resources disseminated online. This is commonly referred to as the digital divide (Norris, 2001). The digital divide often creates a knowledge divide because of the close link between knowledge and technology in society (Unesco World Report, 2005). Also, the privatization of knowledge () allows organizations and institutions to hoard and monopolize acquired knowledge and deny access to the general public regardless of the potential necessity of this knowledge to the general public. For instance, The problems of access are even more pronounced when considering information access within developing countries where less technology and less advanced technology is available. ICT development in South Africa is an excellent example of a developing nation which struggles with extremely low and uneven distribution of ICT technology. This has, in turn, cut them off from knowledge creation opportunities needed to be competitive within global marketplace and reduce poverty. The links between technology, knowledge, and economic growth become sadly apparent in this and other similar cases.

Fifth, there is also inequality in terms of how scientific and technological knowledge is developed around the globe. This is important because of the global impact exerted by many new scientific and technological innovations. Developing countries do not have the same opportunities as developed countries to invest in costly large-scale research and expensive research facilities and instrumentation. This is particularly salient in high tech industries where research and development costs are high and specialized expertise is difficult to obtain. This inhibits developing nations to meet the conditions required for successful scientific and technological innovation. Compounding this is the phenomenon of 'brain drain' where the global interest in scientific and technological expertise accentuates the inequality between developed and developing nations by attracting young talented professionals from the developing and politically torn countries at the expense of the countries of origin (Harrison, 2007).

Sixth, the negative impacts of many scientific and technological innovations have on humans and the environment has led to some scepticism and resistance to increasing dependence on technology within the knowledge society. Doucet's *(2007) Urban Meltdown: Cities, Climate Change and Politics as Usual* examines environmental problems in urban living such as, air pollution, climate change, urban sprawl and transportation. Despite the many advantages of urban living, urban centers are responsible for a great deal of environmental destruction including over 80% of the world's greenhouse gases. Doucet call for city empowerment to have the courage and foresight to make decisions that are acceptable its habitants rather that succumb to global consumer capitalism and the forces of international corporations on national and local governments. In Canada, lucrative mining of asbestos in Tetford Mines (Quebec) for manufacturing and building has exposed workers (and local residents) to highly toxic asbestos fibers linked serious health issues (Quebec Institute of Public Health, 2004). Despite evidence of potential environmental and human dangers, the Quebec trade union movement continues to the asbestos industry and keep asbestos mines open in the province. It is noteworthy that many of the worst environmental impacts are created by the activities of organizations in developed nations within developing nations. On the one hand, outsourcing manufacturing and labouring costs to developing nations allows organizations to cut costs and provide much needed employment to areas of the world in economic strife. On the other hand, numerous examples of environmental destruction, worker exploitation, and unethical business practices threaten the viability of long-term economical growth in developing nations, such as, sweat shops and environmental damage caused in developing countries (Radin & Calkins, 2006; Arnold & Hartman, 2005).

Seventh, scientific and technological innovations that have transformed organizational life within a global economy have also, in many cases, supplanted human autonomy and control in work within a technologically oriented workplace. The supplanting of human with technological driven work has contributed to schisms in human work relations including, a sense of alienation from the work process, unstable tenure and unsatisfactory working conditions (Gossett, 2001), and exclusion from decision making processes in what has been referred to as the age of the disposable worker (Conrad and Poole, 1997). and what it describes is believed to be a common challenge in many work environments, giving rise to strong social and ethical implications.

Eighth, the rapidly changing landscape of organizational life and recent history of unethical business practices has given rise to public debates concerning organizational responsibility and trust. This is salient in the critiques of large organizations and organizational alliances that form the bedrock of the knowledge society. Major organizational scandals from major corporations (e.g., Coca Cola, Enron, Mitsubishi, WorldCom, and Tyco) attest to the growing need for a dialogue on organizational ethics at a global level. This situation is polarized in the technologically advanced context of post-9/11 with heightened concerns revolving around security, international standards, and codes of conduct (May, 2006). The advent of virtual organizations and telework has bolstered ethical problems by providing more opportunities for fraudulent behaviour, deception, and the production of misinformation. Unlike academic research which involved established peer review to ensure quality control and reliability, knowledge production in other areas of work does not have such mechanisms in place. This problem is accentuated in large-scale virtual organizations where a high degree of knowledge fragmentation can result from divisions of labour and increasing specialization in highly technical areas (David & Foray, 2002). For these reasons, concerted efforts are required to uphold ethical values in advancing new knowledge and tools within societal relations which do not exclude people or limit liberties of some people at the expense of others.

NEW DIRECTIONS AND OPPORTUNITIES

The Rise of the Knowledge Society

The specific nature of a knowledge society is constituted by the interaction of its elements (people, knowledge and tools, and relations). Knowledge and tools are a key component in the knowledge society affecting the types of relations and people contributing to its evolution. Powerful new information and communication technologies (knowledge and tools) allow members (people) in a knowledge society to enhance information exchange and cultivate interaction between people and groups of people (relations) previously separated in time and space. The increasing flexibility of mobile ICT's combined with the advent of social networking tools have created many new opportunities for information sharing and knowledge building not possible 20 years ago.

Innovation in science and technology is a key area of interest in most (if not all) knowledge societies and an important component of knowledge-based economies in developed and developing countries around the world. To this end, governmental relations have encouraged key groups of people (academics, scientists, industry leaders) to cultivate relations (funded initiatives, multidisciplinary research groups, programs) to advance scientific and technological innovation. In Canada, for instance, a number of initiatives and groups have been created to increase the role of academic expertise within private sector scientific and technological innovation. The Council of Canadian Academics was created by the federal government to pull together private sector and academic experts to provide expert knowledge to contribute to informed public discussion and decision-making (Council of Canadian Academics, 2007). The Canada Foundation for Innovation (CFI) is another key initiative by the Government of Canada to fund to strengthen Canada's capacity for innovation. Since its creation in 1997, it has invested more than $3.75 billion in support of Canada research projects at 128 research institutions in 64 municipalities across Canada (Canada Foundation for Innovation, 2007). The Grants Program for the Centres of Excellence in Commercialization and Research (CECR) provides support to research and commercialization centres that help connect people, services, and research infrastructure to "position Canada at the forefront of breakthrough innovations in priority areas" (Networks of Centres of Excellence, 2007). This is but a fraction of total investment of other countries with a much larger budget for science and technology innovation (like the United States). These initiatives provide a glimpse of the types of relations and people that come together for knowledge creation and exchange within an evolving knowledge society.

Within a knowledge society, there is opportunity for both ethical and unethical conduct. As will be demonstrated throughout this book, the ethical dimension of a knowledge society is deeply connected to its key elements in pivotal ways that affect its evolution. The types of knowledge and tools developed in society are connected to the types of relations cultivated by members of a society. For instance, advanced areas of science and technology like nanotechnology stimulate relations between scientists and scholars in speciality areas. These relations, if sustained, can become entrenched within specialized knowledge-based communities and institutions within society. This may, in the same instance, limit time and resources available for cultivating different types of relations and communities within society. Due to the increasing complexity of speciality knowledge and tools, this may also limit the type of people that can and cannot participate in knowledge-based communities and institutions within society.

Moreover, the advancement of the knowledge society within an advancing technological context has created a number of opportunities that can help overcome the many new social and ethical challenges. Many of these opportunities are not new, but rather, represent the rediscovery of values and interests that the public fears may be threatened in this new age of scientific and technological innovation. These can be divided into the following key areas: The rise of the knowledge dimension, the rediscovery of identity and community, the rediscovery of organizational trust and accountability, and increased attention to new technologies available for leveraging knowledge building opportunities.

The Rise of the Knowledge Dimension

There is now a greater awareness that information is not enough. There is far more information available than people can process. In a technological advanced society, advanced ICTS increase the amount of information available at any one time which exceeds individuals' capacities to absorb this information, let alone act on this information in a meaningful way. This helps explain why individuals can relate to concepts like information overload and information pollution.

The rise of the knowledge dimension describes the context of society with information overload and a need for new strategies to select what is most valuable.

Knowledge is not information and it is the quality of information that informs real life practices that is important to consider. Not surprisingly, there has been a shift in understanding that knowledge is at the heart of work in a knowledge based economy (Nonaka, Umemoto, & Senoo, 1996). Information consists of collected data about something independent of its possessor. Information is neutral and does not depend on individual cognitive processes to exist. For example, car owner manuals contains gathered facts about cars and their basic functions. Knowledge refers to the acquisition of information through cognitive processes which allow individuals to act in meaningful ways. A mechanic's knowledge of cars is derived from cognitive processes and experiences of cars which allow the mechanic to act in meaningful ways (e.g., fixing a broken car part). In this sense, information is only part of what is needed for knowledge because knowledge requires the internalization of information through limited cognitive capacities

Because of the increased attention to the need for placing knowledge at the centre of inquiry, a variety of new terms and concepts have emerged to concretize the role of knowledge in organizations and society. Some key concepts using knowledge as an organizing construct include, knowledge organization, knowledge management (Bartlett, 1999), knowledge services (Simard, Broome, Drury, Haddon, O'Neil, & Pasho, 2007), tacit and explicit knowledge (Polanyi, 1966), knowledge building (Scardamalia, & Bereiter, 2003), knowledge creation (Nonaka, & Takeuchi,1995), and knowledge revolution(Sakaiya,1991) and knowledge building environments (Scardamalia & Bereiter, 2003). Applications within these concepts draw attention to the importance knowledge based approaches in extending the limits of education and work. Although there are many aspects of acquired knowledge that can be studied, this book highlights ethical aspects of acquired knowledge and how it relates to human activity within a technological society.

The Rediscovery of Identity and Community

There is a growing body of scholarship that delves into current conceptions of identity (i.e., cultural fragmentation, identity crisis, and the de-centering of the subject). This work has helped promote greater awareness of challenges and opportunities arising within a modern technological landscape (Giddens, 1991; Lyotard, 1984*).* How identity is formed and shared in a context of advanced ICT's is a growing concern in society today *(Turkle, 1995)*. As touched on above, *Turkle's (1995)Life on the Screen: Identity in the Age of the Internet* explores how computers and the Internet have changed traditional assumptions about identity and the self from being centered self to a more fragmented and de-centered self interacting in compartmentalized ways within multiple online environments.

There has also been increased attention on how to build identity and community. For instance, Doheny-Farina's (1996) discuss how to reorient thinking from global to local contexts and limit the use technology to create wired neighborhoods that leverage community experiences and anchor community identity. Doheny-Farina's (1996)*The Wired Neighborhood* explores ways to build community and meaning at a local level, "If you want to enhance the culture, steer your participation in the net toward ways that better integrate you and others into your local geophysical communities (p.123)."

Within the organizational context, there has also been a gradual shift to community based approaches for knowledge generation as is reflected in recent work in communities of practice (Argyris, 1993) knowledge building teams (Krogh, Ichijo, & Nonaka, 2000; Nonaka &Takeuchi, 1995) and virtual knowledge building communities (Luppicini, 2007). As discussed by Luppicini (2007), advanced ICT provide highly flexible virtual knowledge building opportunities permitting members to explore shared topics of interest while constructing communal databases of knowledge that can be accessed by all members. Improved workflow and advanced information generation are among the major benefits of using a community based approach to knowledge building.

The Rediscovery of Organizational Trust and Accountability

One of the major outcomes of documented unethical and often illegal practices over the last 20 years is the development of a public awareness of unethical conduct. One of the positive outcomes is that there is increased attention to the much neglected social and ethical dimensions of organizational work within society. This, in turn, has led to the advancement of a rich body of scholarship to address the ethical aspects of organizational life in this evolving knowledge based economy (Cheney, 2004; Dienhart, 2000; Donaldson & Werhane, 1999; Gini, 2005; Johannesen, 1996). This has, in turn, helped reveal a significant gap in knowledge about organizational work that has spurred a number of questions that organizational leaders and researchers are now attempting to address: How can organizational loyalty and trust be instilled in a technologically driven workforce? What are the organizational responsibilities in such a context to uphold ethical standards and practices? Increased attention to organizational accountability has also helped extend ethical considerations to environmental impacts, now perceived to be an important facet of organizational accountability. How it is that organizations are able to assess environmental impacts of new scientific and technological innovations is gradually becoming a core component of doing business. As public awareness of environmental impacts and human agency increases, it is likely that more pressure will be placed on organizations to follow policies of ethical business conduct created to protect the environment.

Technologies For Increasing Knowledge Opportunities

The availability of new tools for generating collective intelligence needed in ethical decision making are constantly evolving. Rapid growth in emerging new scientific instrumentation, network technologies and new media, have revolutionized how individuals collect, create, distribute, and acquire knowledge in work and everyday life. This technological growth allows new knowledge practices that were not possible. The development of calculating machines, databases, and other information storage devices allow human cognitive memory capabilities to be augmented though the use of technology that can be used to access and store information on demand (Karpatschof, 1999). Since, McLuhan's (1962)*Understanding Media*, researchers and technology scholars have explored ways that modern media technologies can extend the mind and sensory capacities in the world to allow more knowledge to be acquired and retained. The next generation of prosthetic and smart technologies offer additional opportunities to advance knowledge building in unique ways that were not possible.

Beyond the development of new instrumentation and material technologies, there are also important technological processes and strategies that have helped redefine knowledge work. For instance, Shön's (1983) *The Reflective Practitioner: How Professionals Think in Action* provides conversational techniques for helping professionals access tacit knowledge about their work process. The author focused on articulating a language of design revealed in guided efforts to verbalize knowledge of design processes. Scardamalia and Bereiter (2003) highlights the importance of other knowledge building technologies in the form of explicit rules that help leverage knowledge processes within knowledge building communities. The knowledge building communities are commited to collective goals in productive knowledge work and adhere to guidelines such as designing knowledge building environments which help hidden aspects of participant expertise to be made transparent to all participants.

Taken together, there is reason to be optimistic about the potential of scientific and technological innovation to be harnessed in a way that adds to the value and meaning of life for individuals within a knowledge society. However, like many major innovations in society, there is a need to organize and clarify the deeper meaning, long term effects, and interests of those affected as to minimize unethical practices.

DISCUSSION QUESTIONS

- Why is the term "knowledge society" a more appropriate metaphor than "information society" or network society" for describing technoethical inquiry in society?

- What are the possible ethical dilemmas connected to the present control of key industries by a handful of large firms?
- How do you think multi-national corporations influence technological innovation?
- Why is culture an important consideration in the knowledge society?
- What are the main ethical dilemmas associated with the current state of the knowledge economy in your opinion?

DISCUSSION

The goal of this chapter was to review the coming of the knowledge society and expose the ethical dimensions underlying key mechanisms by which knowledge is created, managed and applied in society. First, it provided a background review of the emerging notion of "knowledge society" from its recent roots in economics to subsequent developments which grew out of other disciplines and fields. Next, it addressed the growing attention to culture and the technological culture entrenched within the evolving knowledge society. Finally, ethical implications and challenges of the knowledge society were identified and a number of opportunities were highlighted including, the rise of the knowledge dimension, the rediscovery of identity and community, the rediscovery of organizational trust and accountability, and the identification of technologies for increasing knowledge opportunities.

REFERENCES

Aarts, E. H., & Encarnacao, J. (Eds.). (2006). *True visions: The emergence of ambient intelligence*. Berlin: Springer-Verlag.

Abramovitz, M., & David, P. (2000). *American macroeconomic growth in the knowledge society*. New York: Dominique Foray.

Anderson, J. A., & Englehardt, E. E. (2001). *The organizational self and ethical conduct*. Fort Worth, TX: Harcourt.

Argyris, C. *(1993)*. Knowledge for action: A guide to overcoming barriers to organizational change. *San Francisco: Jossey-Bass.*

Arnold, D., & Hartman, P. (2005). Beyond sweatshops: Positive deviancy and global labour practices. *Business Ethics . European Review (Chichester, England)*, *14*(3), 206–222.

Aronowitz, S., & DiFazio, W. *(1994)*. The jobless future: Sci-tech and the dogma of work. *Minneapolis: University of Minnesota Press.*

Ashcroft, M. H. *(1989)*. Human memory and cognition. *New York: Harper Collins Publishers.*

Barney, D. (2003) *The network society.* Cambridge, UK: Polity.

Bartlett, C. (1999). *The knowledge-based organization.* New York: Harper & Row.

Beer, S. (1972). *Brain of the firm.* New York: Herder and Herder.

Bell, D. (1973). *The coming of post-industrial society: A venture in social forecasting.* New York: Basic Books.

Breakwell, G., & Fife-Shaw, C. (1988). Ageism and the impact of new technology. *Social Behavior: An international . The Journal of Applied Psychology, 3,* 119–130.

Canada Foundation for Innovation. (2007). Retrieved September 20, 2007, from http://www.innovation.ca/index.cfm.

Casio, W. (2000). Managing a virtual workplace. *The Academy of Management Executive, 14*(3), 81–90.

Castells, M. (2000). *The rise of the network society. The information age: economy, society and culture* (Vol.1). Malden, UK: Blackwell.

Cheney, G. (2004). Bringing ethics in from the margins. *Australian Journal of Communication, 31,* 35–36.

Conrad, P., & Poole, M. (1997). Introduction: Communication and the disposable worker. *Communication Research, 24,* 581–592. doi:10.1177/0093650297024006001

Council of Canadian Academies. (2006). *The State of Science and Technology in Canada.* Retrieved September 20, 2007, from www.scienceadvice.ca/study.html

Cutler, S. (2005). Ageism and technology. *Generations (San Francisco, Calif.), 29*(3), 67–72.

David, P., & Foray, D. (2002). An introduction to the economy of the knowledge society. *International Social Science Journal, 54*(1), 9–24. doi:10.1111/1468-2451.00355

Dienhart, J. W. *(2000)*. Business, institutions, and ethics: A text with cases and readings. *New York: Oxford University Press.*

Doheny-Farina, S. (1996). *The wired neighbourhood.* Stanford, CT: Yale University Press

Donaldson, T., & Werhane, R. H. *(1999).* Ethical issues in business: A philosophical approach. *Upper Saddle River, NJ: Prentice Hall.*

Doucet, C. (2007). *Urban meltdown: Cities, climate change and politics.* Gabriola Island, Canada: New Society Publishers.

Drucker, P. (1959). *Landmarks of tomorrow: A report on the new 'post-modern' world.* New York: Harper & Row

Drucker, P. (1968). *The age of discontinuity.* New York: Harper & Row.

Dyson, E., Gilder, G., Keyworth, G., & Toffler, A. (1994). *Cyberspace and the American dream: A magna carta for the knowledge age.* Future Insight 1.2. The Progress & Freedom Foundation.

Elzinga, A. (1998). Theoretical perspectives: Culture as a resource for technological change. In M. Hard & A. Jamison (eds.) *The intellectual appropriation of technology* (pp.17-32). Cambridge, MA: MIT Press.

Franklin, U. (1990). *The real world of technology.* Concord, MA: House of Anansi Press Limited.

Fukuyama, F. (2007). *Our posthuman future.* Munich: Holtzbrinck.

Garnham, N. (2004). Information society: Theory as ideology. In F. Webster (Ed.) *The information society reader.* London: Routledge.

Giddens, A. (1991) *Modernity and self identity.* Cambridge, UK: Polity Press.

Gini, A. (2005). *Case studies in business ethics* (2nd ed.). Upper Saddle River, NJ: Prentice Hall.

Gossett, L. (2001). The long-term impact of short-term workers: The work life imposed by the growth of the contingent workforce. *Management Communication Quarterly, 12*(1), 8–27.

Haag, S., Cummings, M., McCubbrey, D., Pinsonneault, A., & Donovan, R. (2006). *Management information systems for the information age (3rd Canadian Ed.).* Canada: McGraw Hill Ryerson.

Haraway, D. (1991). *A cyborg manifesto, science technology and socialist-feminism in the late twentieth century, the reinvention of nature.* New York: Routeledge.

Hardt, M., & Negri, A. (2000). *Empire.* Cambridge, MA: Harvard University Press.

Harrison, F. (January 8, 2007). Huge cost of Iranian brain drain. *BBC News.* Retrieved March 1, 2009, from http://news.bbc.co.uk/1/hi/world/middle_east/6240287.stm

Jenkins, H. (2006). *Convergence Culture: Where Old and New Media Converge.* New York: NYU Press.

Johannesen, R. J. (1996). *Ethics in human communication,* (4th ed.). Prospect Heights, IL: Waveland Press.

Kahin, B., & Foray, D. (2006). *Advancing knowledge and the knowledge economy.* Cambridge, MA: MIT Press.

Karpatschof, B. (1999). The meeting place of cognition and technology. In J. P. Marsh, J. L. Mey, & B. Gorayska (Eds.), *Humane interfaces: questions of method and practice in cognitive technology* (pp. 115-125). Amsterdam: Elsevier.

Knorr-Cetina, K. (1999). *Epistemic cultures: How the sciences make knowledge.* Cambridge, MA: Harvard University Press.

Krogh, G., Ichijo, K., & Nonaka, I. (2000). *Enabling knowledge creation: How to unlock the mystery of tacit knowledge and release the power of innovation.* Oxford, UK: Oxford University Press.

Latour, B., & Woolgar, S. (1986). *Laboratory life: The construction of scientific facts.* Princeton, NJ: Princeton University Press.

Luppicini, R. (Ed.). (2007). *Online learning communities.* Greenwich, CT: Information Age Publishing.

Lyotard, J. (1984). *The postmodern condition.* Manchester, UK: Manchester University Press.

Machlup, F. (1962). *The production and distribution of knowledge in the United States.* Princeton, NJ: Princeton University Press.

May, S. (Ed.). (2006). *Case studies in organizational communication: Ethical perspectives and practices (pp. 75-85).* Thousand Oaks, CA: Sage.

McCann, R., & Giles, H. (2002). Ageism and the workplace: A communication perspective. In Nelson, T. (Ed.), *Ageism* (pp. 163-199). Cambridge, MA: MIT Press.

McLuhan, M. (1962). *Understanding media.* London: Routledge.

McLuhan, M. (1964). *Understanding media: The extensions of man.* Toronto: McGraw Hill.

Middleton, H. (2005). Creative thinking, values and design and technology education. *International Journal of Technology and Design Education, 15*(1), 61–71. doi:10.1007/s10798-004-6199-y

Mills, J. S. (1831). *The spirit of the age.* London: WW Norton.

Networks of Centres of Excellence. (2007). Retrieved September 20, 2007, from www.scienceadvice.ca/study.html.

Nonaka, I., & Takeuchi, H. (1995). *The knowledge-creating company.* New York: Oxford University Press.

Nonaka, I., Umemoto, K., & Senoo, D. (1996). From information processing to knowledge creation: A paradigm shift in business management. *Technology in Society, 18*(2), 203–218. doi:10.1016/0160-791X(96)00001-2

Polanyi, M. (1966). *The tacit dimension.* New York: Doubleday & Co.

Quebec Institute of Public Health. (2004). *The Epidemiology of Asbestos-Related Diseases in Quebec.* Retrieved, July 15, 2008 from www.inspq.qc.ca cote: IN-SPQ-2004-033.

Radin, T., & Calkins, M. (2006). The struggle against sweatshops: Moving toward responsible global business. *Journal of Business Ethics, 66*(2-3), 261. doi:10.1007/s10551-005-5597-8

Real, M. R. (1975). Cultural Studies and Mediated Reality. *Journal of Popular Culture, 9*(2), 81–85.

Robertson, R. (1992). *Globalization: Social theory and global culture.* London: Sage.

Rooney, D., Hearn, G., & Ninan, A. (2005). *Handbook on the knowledge economy.* Cheltenham, UK: Edward Elgar.

Sakaiya, T. 1991. *The knowledge-value revolution.* New York: Kodansha International.

Scardamalia, M. (2003). Knowledge Society Network (KSN): Toward an expert society for democratizing knowledge. *Journal of Distance Education, 17*(3), 63–66.

Scardamalia, M., & Bereiter, C. (2003). Knowledge building environments: Extending the limits of the possible in education and knowledge work. In A. DiStefano, K. E. Rudestam, & R. Silverman (Eds.), *Encyclopedia of distributed learning (pp. 269-272)*. Thousand Oaks, CA: Sage Publications.

Schön, D. (1983). *The reflective practitioner: How professionals think in action.* Basic Books: New York.

Simard, A., Broome, J., Drury, M., Haddon, B., O'Neil, B., & Pasho. D. (2007). *Understanding knowledge services at natural resources Canada.* Natural Resources Canada, Office of the Chief Scientist, Ottawa.

Stehr, N. (2002) *Knowledge & Economic Conduct.* Toronto: University of Toronto Press.

Tichi, C. (1987). *Shifting gears: Technology, literature, culture in modernist America.* Chapel Hill, NC: University of North Carolina Press

Tomlinson, J. (1999). *Globalization and culture.* Chicago: University of Chicago Press

Turkle, S. (1995). *Life on the screen: Identity in the age of the Internet.* Toronto: Simon and Schuster.

Turnage, J. (1990). The challenge of new workplace technology for psychology. *The American Psychologist, 45,* 171–178. doi:10.1037/0003-066X.45.2.171

Williams, R. (1975). *Television: Technology and Cultural Form.* New York: Schocken.

World Report, U. N. E. S. C. O. (2005). *Towards Knowledge Societies.* Retrieved March 1, 2009, from http://unesdoc.unesco.org/images/0014/001418/141843e.pdf

ADDITIONAL READING

Bell, D. (1976). *Post-industrial society.* New York: Basic Books.

Bunge, M. (1977). Towards a technoethics . *The Monist, 60*(1), 96–107.

Burck, G. (1964, November). Knowledge: the biggest growth industry of them all. *Fortune,* 128–131.

Machlup, F. (1980). *Knowledge: Its creation, distribution, and economic significance.* Princeton, NJ: Princeton University Press.

Mayer, R. H. (1999). Designing instruction for constructivist learning. In C. M. Reigeluth (Ed.), *Instructional-design theories and models volume II (pp. 141-160).* Mahway, New Jersey: Lawrence Erlbaum Associates, Publishers.

Norris, P. (2001). *Digital divide: Civic engagement, information poverty, and the Internet worldwide.* Cambridge: Cambridge University Press

Saetre, A., & Sornes, J. (2006). Working at home and playing at work: Using ICTs to break down the barriers between home and work. In May, S. (Ed.), *Case studies in organizational communication: ethical perspectives and practices (pp. 75-85).* Thousand Oaks, CA: Sage.

Senge, P. M. (2006). *The fifth discipline: The art and practice of the learning organization.* Toronto: Double Day Publishing.

Snoeyenbos, M., Ameder, R., & Humber, J. (2001). *Business ethics* (3rd ed.). Amherst, MA: Prometheus Books.

Turkle, S. (2005). *The second self: computers and the human spirit, twentieth anniversary edition.* Boston: MIT Press.

Wenger, E. (1998). *Communities of practice: Learning, meaning, and identity.* Cambridge: Cambridge University Press

Wenger, E., McDermott, R., & Snyder, W. (2002). *Cultivating communities of practice.* Boston: Harvard University Press.

Chapter 2
Technoethics

INTRODUCTION

Being an affectionate parent does not exculpate any crimes; being a competent engineer does not confer rights of piracy on the environment; being an efficient manager does not entitle him to oppress others. Every human being has a number of intertwined responsibilities and each of them is as personal and intransferable as a joy or grief. --Bunge, 1977, p. 96.

A major struggle within our evolving knowledge society is that increasingly potent scientific and technological growth is forcing individuals to re-examine how technology is viewed. This is especially salient in the pure and applied sciences where technological developments offer ways to surpass current human capacities and

DOI: 10.4018/978-1-60566-952-6.ch002

affect life in ways that were not imaginable fifty years ago. New breakthroughs in medicine, information and communication technology, transportation and industry are juxtaposed with growing needs to deal with moral and ethical dilemmas associated with new technological developments. Increased reliance on new technology creates fundamental challenges revolving around security and privacy issues, access issues to education and health care, legal issues in online fraud and theft, employer and government surveillance, policies issues in creating and implementing ethical guidelines and professional codes of conduct, along with ethical dilemmas in a number of vital areas of research and development.

Juxtaposed with the coming of powerful new scientific and technological advances, is a major push to rediscover the ethical dimension of technology across the sciences, social sciences, and humanities. Literature on ethics and technology are in abundance which focus on key areas of technology and ethics; *Ethics and Technology: Ethical Issues in an Age of Information and Communication Technology* (Tavani, 2007), *On Technology, Medicine and Ethics* (Jonas, 1985), *Information Ethics* (Floridi, 1999), *and Computer Ethics* (Johnson, 1985). However, there is no authored book on ethics and technology dedicated to the diverse areas of research and theory in use today.

Surmounting debates and scholarly inquiry across multiple disciplines and applied fields of study connected to technology form the basis of an emerging field known as Technoethics (TE). Technoethics (TE) is concerned with all social and ethical aspects of the design, development, utilization, management, and evaluation of science and technology in society. As an interdisciplinary field, it utilizes theories and methods from multiple knowledge domains to provide insights on ethical dimensions of technological systems and practices for advancing a technological society. Technoethics provides a systems theory and methodology to guide a variety of separate areas of inquiry into technology and ethics (see Chapter 5). As it is developed in this text (see Chapter 4), technoethical inquiry is a systems theory and methodology for guiding technological systems research and practice in key areas of human-technological activity (I.e, technology assessment, technology design).

As will be discussed in this chapter, Technoethics derives from a longstanding history of scholarship in ethics, applied ethics, philosophy of technology, and technocritical writing in the humanities. The emergence of Technoethics as a formal field can be connected to a marriage of scholarship on technology and ethics that spread from specialty areas within philosophy to other disciplines concerned with social and ethical issues regarding technology. This marriage placed social and ethical questions about technology on new ground and provided new opportunities that crossed disciplinary boundaries. This eventually led to the emergence of Technoethics as a field of interdisciplinary scholarship to help ground a number of areas of technoethical inquiry that previously were separate and isolated programs.

BACKGROUND

Ethics and Ethical Theories

The Oxford Concise Dictionary (2006) states, "Ethics are the moral principles influencing conduct" (p. 490). In this sense, ethical theories are embedded in how people reason about what they should do and how they should conduct themselves. More specifically, Ethics constitutes a branch of philosophy concentrating on questions of moral conduct and good life. Ethics describe moral principles influencing conduct and the study of ethics is rooted in what people do and how they believe they should act in the world. It focuses on values and actions of individuals within society and can be traced back to Socrates and the importance he placed on cultivating self-knowledge (know thyself). He believed that the cultivation of self-knowledge (the highest form of knowledge) will guide individuals to do naturally do what is good and right (Plato,). Since this time, the academic pursuit of Ethics has given rise to a diverse array of ethical theories and concepts. Ethical theories are theories about moral standards and how these standards guide conduct. Ethical theories address foundational questions of responsibility and accountability, right and wrong, and good and evil. While ethical perspectives derived from descriptive ethics (What does an individual believes is right?) and applied ethics (How can an individual derive moral knowledge and put it into practice?) are becoming important, the bulk of ethical theories available are normative theories (How should an individual act?). Normative theories of ethics may focus on moral character (virtues), duties or rules (deontology), relationships (relations), or the consequences of actions (consequentialism). Major ethical theories discussed today include: Utilitarianism, duty ethics, rights ethics, virtue ethics, and relational ethics.

Utilitarian ethical theories are based on the assumption of the greatest happiness principle aligning good with that which brings the greatest happiness to the greatest number of people. Utilitarianism is concerned with consequences and ethical value discerned from the outcome of actions as a means to an end, rather than focusing on intentions or virtues. Ethical actions are assessed according to their positive outcomes relative to alternatives. Utilitarian ethics is traced to Jeremy Bentham's argument that pain and pleasure were the only intrinsic values in the world. John Stuart Mills extended this theory in highlighting ethical actions as those that provide the greatest happiness for the greatest number of people. This formed the basis of one of two major variations of utlilitarianism, act utilitarianism. Act utilitarianism describes ethically right actions as those that yield the greatest amount of happiness for the greatest number of people. Under rule utilitarianism, ethically right actions are those that follow ethical principles oriented to creating the most happiness. Typical questions of utility might include: What actions or rules will produce maximum good and minimal harm for the greatest number of people? Have all possible actions and alternatives for maximizing the greatest good been considered?

Duty ethics is generally concerned with the obligations one has to others in society. These duties often expressed as universal, rational rules that are self evident (e.g., Do unto others as you would be done by). Duty ethics is a deontological theory of ethics since it focuses on the rightness of actions themselves, rather than the consequences of those actions. It is designed to explain what individuals ought to do in order to live an ethical life. Kant held the paramount deontological theory of ethics, which he framed around the notion of the categorical imperative which stated, "Act only according to that maxim whereby you can at the same time will that it should become a universal law" (Kant, 1785). From this Kant reasoned that individuals should act in such a way that you always treat all individuals as ends in themselves, rather than mere means to ends. He thought that individuals were capable of coming to agreement on what benefits everyone through reason. For Kant, reason was the core to ethical decision making because it guides individual to act from duty rather than from personal interest. A good action for Kant is one that done by free will in accordance with reason. Ethical decision making for Kant focuses on ethical actions (not the consequences of these actions) guided by good will and the respect for others as rational beings with intrinsic values, not as means to ends. Typical questions of duty might include: What actions are inherently right without qualification? What actions are inherently good without considering their affect on others? What ethical principles should any individual be willing to follow that would become a rule of conduct that everyone in society would follow?

A rights perspective is generally concerned with the obligations between individuals based on assumptions of rights owed to individuals by the collective they are part of. A rights perspective views the rights of individuals to be universal for all individuals, natural, and self-evident. Social contracts are used to ensure that individual rights are upheld in society. This approach to ethics via contractual agreements of acceptable and unacceptable conduct is rooted in natural law and the writings of Thomas Hobbes. Hobbes (1651) *Leviathan* provided the base for rights ethics that led to later formulations by John Locke and Rawls. Hobbes viewed ethical rights as natural rights of all individuals in society closely linked to obeying accepted political authority. John Locke viewed ethics as connected to human nature as rational and tolerant within a natural state where all people were created equal and independent, and none had a right to harm another or take their rights away (Locke, 1690). Locke believed that the judgement of rights defined within a social contract applied to the conduct of individuals and the conduct of institutions. He promoted the need for governmental checks and balances to ensure that individual rights were satisfied. In a slightly different vein, John Rawls believed that ethical conduct could best be decided through adopting an original position where individuals assume limited knowledge of the past, present, and future in order to avoid biases in judgement that overemphasize anticipated personal harm and benefits over that of others (veil of ignorance). John Rawls' (1970)*Theory of Justice*, pulls together concerns about liberty and equality and provides an ethical perspective of justice as fairness. The justice as fairness perspective stipulates that all individuals have the right to basic liberties (liberty principle) and that social and economic inequalities should benefit everyone and be organized in positions open to all (difference principle). Typical questions of rights might include: How can the human rights and dignity of all individuals and groups be maintained in society? What policies and procedures are just and unjust in society? How can diverse rights be preserved within society? What is the current held social contract in society and how can it be improved? How can individuals and groups of individuals conduct themselves as to not give unfair advantages to some at the expense of others?

Virtue Ethics can be traced to the work of Plato and Aristotle within ancient Greek philosophy (and areas of Chinese Philosophy). It views human character and efforts to lead the good life as core in ethical thinking and conduct. Under this perspective, all individuals are born with inherent potential (virtues) that must be developed and self-actualized in life and society. Ethical actions are those that allow for the full expression of human potential (behaviours and habits) that creates benefits held to be important to society. For Plato, leading the good life could be lived by cultivating virtues, notably, courage, wisdom, temperance, and justice. Plato believed that social and political life should be based on absolute moral truths that every individual would follow. Highlighting the importance of political and social life was reinforced through Aristotle's view of virtues and conception of humans as social animals. In *Nicomachean Ethics*, Aristotle described the development of virtue as taking place in the social sphere through the obligations of contributing members of society. Society, in turn, is obligated to provide individuals opportunities to develop their full potential. Key modern proponents of virtue ethics include Philippa Foot (1978) and others. Typical questions of virtue might include: What is the nature of justice? What is the most virtuous way to conduct oneself in society? What strategies for individuals and groups of individuals can be offered to help cultivate harmonious and happy lives in society? How can individuals act to self-actualize and develop their full potential in society?

Relationship ethics is a relatively recent variety of normative theories of ethics . It is concerned with the ethics of care and consideration as derived from human communication. Relationship ethics views ethical communication as the core of healthy relationships and the ground upon which ethical actions are built. The relationship perspective focuses on the processes which promote the attention to and ongoing care of productive relationships among individuals and groups within society. Key proponents include Martin Buber, Mikhail Bakhtin, and Carol Giligan. In *I and*

Thou, Martin Buber addressed the importance of communicating with others as fundamental in allowing for the development of personality and self in society. He advocated the use of perspective taking in communication as a way of developing an appreciation of others' views and experiences as well as contributing to a sense of social responsibility concerning the potential influence of ones' actions on others. Mikhail Bakhtin contributed to relationship ethics an appreciation of relationship dynamics and ethics. He focused on ethical actions in terms of how communicative acts can promote change (centripetal forces) or stability (centrifugal forces) in relationships. Another development in relationship ethics is the ethics of care derived from Carol Giligan. Based in a feminist framework, she views mutuality and reciprocity as core principles for guiding ethical actions to maintain caring relationships between individuals within society. Typical questions of relationship might include: How can individuals leverage relationships to benefit all individuals? How can individuals care for others through communication? What principles can be used to help maintain ethical relationships between individuals within society?

Ethical Tensions and the Turn to Applied Ethics

Which ethical theory is right and which one should people follow? The problem with relying on ethical theories as a stand alone approach for dealing with social and ethical issues is that ethical theories are based on different assumptions about the nature of reality and the obligations of individuals within it. Ethical tensions arise when the fundamental assumptions underlying different ethical theories leads to different conclusions concerning what constitutes ethical conduct. Following the complete prescriptions of any one ethical theory would lead to violations of other ethical theories. For instance, what if a situation arose where saving the lives of multiple individuals depended on the death of one, deontological ethics would disallow any action that would sacrifice a human life to save multiple lives since each life must be respected as an end in itself. Robert Nozick referred to this as the paradox of deontology (Nozick, 1974). Under utilitarian ethics, such actions would be permitted in order to maximize benefits and minimize harm for the many. This, however, leads to debate about how human life is valued since utilitarianism undermines the value of humankind by ignoring personal intuition and reducing individuals to means for achieving desired consequences. Another example of ethical tension arises when looking at controversial societal practices, such as slavery. Under a rights perspective, enslaving individuals is unjust because it undermines equality and denies people fundamental rights. A traditional virtue ethics perspective struggles with such practices and behaviors, since in some cultures and historical periods, slavery was not considered an ethical problem and in others it was. Cultural relativists argue that different cultures and societies can have diverging views on what constitutes a virtue that virtue ethics does not accommodate.

The emergence of a new form of ethical inquiry, referred to as Applied Ethics, provided an alternative ethical framework to traditional Ethics. Applied ethics developed as branch of philosophy focused on the application of ethical theories to real world events and issues. A variety of important topics affecting human life are addressed within various sub-fields of Applied Ethics, including, Medical Ethics, Bioethics, Business Ethics, Environmental Ethics, Legal Ethics, Marketing Ethics, International Ethics, Public Administration Ethics, human rights issues, and animal rights issues. Peter Singer's (1986)*Applied Ethics* provided a solid grounding for the study of Applied Ethics that led to a diversification of issues and concepts found in later works such as the *Encyclopedia of Applied Ethics* (Chadwick, 1997). Typical questions raised in Applied Ethics address real life problems that affect people in society: Is it acceptable to give advantages to members of marginalized groups to equal the playing field? What ethical obligations (if any) do individuals from of rich countries have to help poor countries? To what extent (if any) should people be responsible to protect and preserve the environment? What ethical responsibilities do people have (if any) to uphold animal rights?

To conclude, following the longstanding tradition in Ethics, the rise of Applied Ethics marked a departure from traditional Ethics by bringing philosophical inquiry into the realm of real world human problems, events, and practices. This anticipated the coming of specialized areas of human activity (i.e., the pursuit of science and technology) with new ethical dilemmas that eventually led to what would become known as Technoethics. While this section traced out developments in Ethics and Applied Ethics, the next section focuses on other historical-philosophical developments underlying Technoethics rooted in Pragmatism and the Philosophy of Science and Technology.

TECHNOLOGICAL INQUIRY AND THE EARLY GROUNDWORK FOR TECHNOETHICS

Institutionalizing Technological Inquiry

Changes in societal views about the importance of technology in the 1960s and 1970s mirrored in scholarly work on technology, helped advance public understanding about the need to study science and technology from a variety of perspectives and standpoints. This led to a greater institutionalization of science and technology within professional organizations and educational institutions. Professional associations (i.e., The Society for Philosophy and Technology, the Society for the History of Technology, and the Society for Social Studies of Science) helped nurture technological inquiry by attracting philosophers, historians, social scientists, and other scholars with technology research interests and by sponsoring journals (e.g., *Techné: Research in Philosophy and Technology, Science, Technology, and Human Values, Technology and Culture*) focused on the social study of science and technology in a variety of areas. In 1958, the Society for the History of Technology (SHOT), was created as a professional society for historians of technology. It hosts academic conferences and publishes the journal, Technology and Culture. In 1975, the Society for Social Studies of Science was established. It hosts regular conferences and publishes the journal, *Science, Technology, and Human Value*. With a more philosophical orientation, the Society for Philosophy and Technology has existed since 1976 and promotes philosophically studies of technology through hosting conferences and the publication of *Techné: Research in Philosophy and Technology*.

In the 1980s, a major contributor to the institutionalization of science and technology is linked to the rise of Science and Technology Studies (STS). One main STS orientation is based in Philosophy. The branch of STS based in Philosophy is highly theoretical in orientation and derives from history and philosophy of science (Fuller, 2005). A second STS orientation is based in the Social Sciences and has more of an applied focus and a central aim of making science and technology useful to society. This second orientation was driven largely by developments in Sociology and the study of scientific knowledge, which eventually expanded in scope to focus on science and technology (Woolgar, 1991).

University degree programs in Science and Technology Studies (STS) provided instruction and research opportunities to advance the study of science and technology. The development of Science and Technology Studies programs were intended to leverage understanding and inform decision-making revolving around scientific and technological innovation. These programs focus on key questions of technological interest such as, How do social, political, and cultural values affect the advancement of scientific and technical innovations and how do scientific and technical innovations affect society, politics, and culture? This also set the stage for the emergence of Technoethics through the growth of sophisticated philosophical and theoretical analyses of technology.

Philosophy of Technology

Perhaps more than any other academic area, mid 20th century, Philosophy of Technology provided crucial grounding for Technoethics by formalizing philosophical inquiry about technology and society. Martin Heidegger's *The Question Concerning Technology* was a pioneering work in that it delved into the new connections to the world made possible by modern technology (Heidegger, 1977). Heidegger considered this new technological connection between human beings and the world to be problematic because it assumed that individuals and other living entities were objects capable of being manipulated. A less pessimistic interpretation of technology as connected to human activity derives from McGinn (1978). McGinn's conceptualized technology as a value-laden human activity linked to socio-cultural and environmental influences. McGinn (1978) helped advance the view that technology was value-laden in terms of its close connection to the values of those who create and use it, its monopolization of limited resources at the expense of other possibilities, and in terms of individual and cultural values of designers expressed in technological products. Other seminal writing contributing to the growth of Philosophy of Technology include: Contributions to a Philosophy of Technology (Rapp, 1974), *Autonomous Technology: Technics-out-of-Control as a Theme in Political Thought* (Winner, 1977), Technics and Praxis (Ihde, 1979); *Technology and the Character of Contemporary Life: A Philosophical Inquiry* (Borgmann,1984*), Philosophy of Technology* (Ferré, 1988);*Thinking Through Technology* (Mitcham, 1994), *Questioning Technology* (Feenberg, 1999), *Technology and the Good Life* (Higgs, Light & Strong, 2000), *American Philosophy of Technology* (Achterhuis, 2001) and *Readings in the Philosophy of Technology* (Kaplin, 2004). This body of scholarship helped legitimize philosophical inquiry into technology as it bears on human life and society. Although the core focus in the Philosophy of Technology was mainly on technology, this body of work provided grounding for other scholars invested in advancing philosophical inquiry in technology and ethics (Jonas, 1985).

Theories of Technology and the Technocritical Turn

What can theories of technology and technocritical studies contribute to our understanding of technology and ethics in society? In the 1970s and 1980s, an amassing body of critical studies about technological developments became available, partly inspired by philosophical critiques of science posited by Thomas Kuhn's famous book *The Structure of Scientific Revolutions* (1962). In this book, Kuhn showed that there are more human values involved in science than suggested by earlier (positivist) philosophy of science. As will be seen, the study of technology within the Applied Sciences, Social Sciences and Humanities (technocritical scholarship) had an enormous impact of how accepted views, aspirations, and fears about technology were changing in the mid and latter part of the 20th century . Technocriticism emerged from critical theory and focused on the study of technological change. Technocriticism expanded the scope of technology studies into an examination of private and public uses of technology, along with the relations among these different uses. This was a new approach to technological inquiry in that it focused on empirical evidence, rather than generalized conceptions of technology as found in the Philosophy of Technology. It was assumed that accurate description of how specific technologies develop could provide a useful resource for critical reflection on the effects of technological developments. Leading figures included Jacques Ellul, George Grant, and Ursula Franklin.

A number of key theories of technology have helped to create intellectual groupings among scholars which have advanced understanding of science and technology studies while providing theoretical frameworks to help explain technological phenomena of interest. Notable theories of technology include technological determinism, social constructivism, cultural and critical theories of technology, and actor-network theory.

Technological determinism was the first substantial theory of technology to receive widespread recognition. It assumes that there are underlying material and physical laws that cause technology to develop in certain ways that are beyond human control. It focuses on technological impacts and how technology shapes human activity and society while raising questions about the possibility of free will and human responsibility when it comes to dealing with technology. Ellul's (1964) *The Technological Society* and Winner's (1977) *Autonomous Technology: Technics-out-of-Control as a Theme in Political Theory* epitomize the spirit of technological determinism. *Jacques Ellul's (1964) The Technological Society* addressed technology from a deterministic standpoint, exploring technological control over humanity and its potential threat to human dignity and freedom. Ellul viewed technology (as technique) as "the totality of methods rationally arrived at and having absolute efficiency (for a given stage of development) in every field of human activity" (p. xxv). This text helped map out some of the potential threats and injustices that could be created in the absence of critical discourse that goes beyond technological efficiency. As highlighted by Ellul (1964), "what is at issue here is evaluating the danger of what might happen to our humanity in the present half-century, and distinguishing between what we want to keep and what we are ready to lose, between what we can welcome as legitimate human development and what we should reject with our last ounce of strength as dehumanization. Ellul's work warns of the dangers in modern technological society to enslave people by systematically impose technology at the expense of human autonomy. He stated, "technique enslaves people, while offering them the mere illusion of freedom, all the while tyrannically conforming them to the demands of the technological society with its complex of artificial operational objectives" (35). In a different area, Winner's text deals with how contemporary conceptions of technology are out of tune with reality and our ability to make informed decisions about technology. Winner (1977) argues that the "contemporary experience of things technological has repeatedly confounded our vision, our expectations, and our capacity to make intelligent judgments" (2). This analysis of technology examines important schisms resulting from technological development, constraining factors exerted by technological growth that limit human control, and historical conditions underlying this rapid growth. Winner noted, "Man has mounted science and is now run away with. I firmly believe that before many centuries more, science will be the master of man. The engines he will have invented will be beyond his strength to control" (44).

A second set of theories of technology fall under the heading of social constructivist theories. Social constructivist stances highlight the importance of social groups and flexibility in technological change process. This stance was largely in opposition to existing determinist stances on technology which highlighted linearity in technological development with a high degree of predictability on outcomes and a focus on technological impact. One notable social constructivist study by Pinch and Bijker (1987) demonstrated how the power of key stakeholder groups influenced the design of bicycles in the late 19th century. The social construction stance explored how various stakeholder groups exerted social influence on developing technology (bicycle design) revolving around issues of interest including dress code (women cyclists), macho appeal (young men cyclists), costs (manufacturers), safety (older cyclists) and speed (sport cyclists). Given a different time and place in history with different types of relations and groups, the development of the bicycle would have been different. Another important work from Latour and Woolgar (1979) explores how everyday work by scientist working a laboratory leads to the social construction of facts. The research shows how facts are constructed in a laboratory setting. This important work documents how scientists in laboratory back up social constructions of facts as a means of leveraging the construction of new facts:

It is not simply that phenomena depend on certain material instrumentation; rather, the phenomena are thoroughly constituted by the material setting of the laboratory. The artificial reality, which participants describe in terms of an objective entity, has in fact been constructed by the use of inscription devices (64).

Taking a slightly different approach to the social study of science and technology, Network theory (ANT) posited by Latour (2005) analyzes social situations in terms of the various types of relationships emerging in connection to technology. Actor-network theory typically attempts to explain how networks of humans (and non-humans) work by focusing on agent interactions which form a whole. One unique aspect of the actor-network theory is that it does not limit agency to human subjects, but allows for non-human agency through associations of humans and nonhumans. Latour argues that "every artifact, such as for example a machine, can be understood only in terms of the meaning which its production and use have had or will have for human action; a meaning which may derive from a relation to exceedingly various purposes" (78). The ANT is a particularly useful development in technological theory that allows for the study of human and non-human relationships within society. This work helped contribute to an understanding of scientific study as socially constructed within the laboratory and connected to the humans that collect and interpret scientific data.

Overall, contemporary philosophy of technology, theories of technology, and technocritical scholarship helped draw attention to the importance of contextualizing technology within societal structures, activities, and practices in the human world. Although this body of work was primarily concerned with how technology influences social order, the ethical considerations raised in this body of reviewed work helped set the stage for the advent of Technoethics in two important ways. First, this work encouraged scholars to invest special attention to key areas of technology in order to ground growing efforts to develop socially acceptable guidelines for technology use in society. Second, this work bridged philosophical inquiry into technology with related work within the applied sciences and Social Sciences. It did so by drawing on real world developments to help situate the discourse on technology within the context of human activity. The outcome of this is that many scholars became more aware of the need to connect technological development to social and ethical aspects entrenched in how people live, work, and play. This placed ethical questions related to technology on new ground requiring special status, which led to the emergence of Technoethics.

THE TECHNOETHICAL TURN

Contemporary Roots of Technoethics

Technoethics was officially defined by Mario Bunge in the 1970s (Bunge, 1977) when arguing for increased moral and social responsibility among technologists and engineers concerning their technological innovations and applications. Bunge (1977) stated, "The technologist must be held not only technically but also morally responsible for whatever he designs or executes: not only should his artifacts be optimally efficient but, far from being harmful, they should be beneficial, and not only in the short run but also in the long term." Bunge recognized the need for creating a new type of ethical inquiry to address the special problems posed by science and technology (Bunge, 1977).

Early efforts to formally define Technoethics provided a much needed unifying framework for a number of independently emerging areas of academic research and study on technology and ethics that preceded any formal definition of "Technoethics". Between the 1890s and 1960s, the precursor to scholarly work in Technoethics was found in the Pragmatism of Dewey and Pearce and the recognition of the value-laden nature of scientific knowledge as well as the emphasis on responsible use of technologies in the face of industrialization with technologies powerful enough to destroy the environment. This was over 50 years before "Technoethics" was officially coined (See Chapter 4 for more details). Moreover, the core branches of contemporary Technoethics (See Chapter 5) grew out of various sub-areas of Philosophy, the Social Sciences, and interdisciplinary domains of academic scholarship on ethics and technology.

One important area of academic scholarship on ethics and technology emerged from Computer and Engineering Ethics. Beginning in the 1960s, Computer and Engineering Ethics contributed early groundwork for Technoethics by focusing on specialized professional codes and responsibilities of engineers concerning their technological creations, which Bunge (1977) used to conceptualize the formal field of Technoethics. Computer and Engineering Ethics was intended to help guide ethical decision-making in areas where technologists applied new technical and scientific knowledge to provide solutions to conflicting societal needs in a variety of domains such as, Aerospace, Agriculture, Bioengineering, Chemical Engineering, Civil Engineering, Construction, Electrical Engineering, Industrial Engineering, Mechanical Engineering, and nuclear energy creation. This was especially important in raising public attention to the effects of technology within natural environment (Passmore, 1974; Rolston III, 1975; Stone, 1972). This early work led scholars to begin considering ethical aspects of urban landscapes and other constructed environments. Jacobs' (1961) *The Death and Life of Great American Cities* was a pioneering study raising Technoethical concerns about American urban renewal policies of the 1950s. Concurrently, a surge of new academic research focused on human aspects of computer use and ethical guidelines inspired by the pioneering work of Norbert Wiener in Cybernetics (the science of control and communication in animals and machines) carried out in the 1940s and 1950s (Wiener, 1948, 1954). This, in turn, provided grounding for the field of Technoethics by highlighting the need to address integrate technoethical consideration in education and professional training (Educational Technoethics) within professions that actively use new technologies such as, Engineering, Journalism, and Medicine.

A second important area of academic scholarship on ethics and technology emerged from Bioethics and Medical Ethics. In the mid-1960s and early 1970s, work in Bioethics and Medical Ethics focused on new technologies leading to the development of Biotech Ethics. This was partly in response to the advent of reproductive medicine and research which raised new ethical issues complicated by new technologies such as, abortion, animal rights, eugenics, euthanasia, *in vitro* fertilization, and reproductive rights. Also at this time, biotechnology (biological technologies) research became increasingly applied to agriculture and food science. Increased use of biotechnologies in the 1980s and 1990s quickly spread to medical research, health care, and industrial applications which gave rise to ethical considerations about cloning, genetically modified food, gene therapy, human genetic engineering, drug production, reprogenetics, stem cell research and other new applications in biotechnology.

A third important area of academic scholarship on ethics and technology emerged as a response to the information revolution and work in information and communication technology (ICT). The establishment of information science and subsequent development of information ethics helped raise concerns about the centrality of information technology and its ethical implications (information ethics). In the 1960s and 1970s, McLuhan spearheaded scholarly work in communication media and media discourse which helped connect communication and technology considerations. Concurrently, information studies pursued in multiple domains (e.g., computer science, library science, cognitive science, media studies, communications, etc.) spurred work in information and communication ethics now recognized as a confluence of the ethical concerns arising from a number of emerging research niche areas (i.e., media ethics, journalism ethics library, computer ethics, cyber ethics). Beginning in the late 1980s and early 1990s, important connections were made between communication and ethics, largely due to communication theorists like Jurgen Habermas, and Benhabib. This helped connect ethical inquiry with information and communication technology applied in the real world. Concurrently, the advent of the Internet and new computing technologies (e.g., spyware, antivirus software, web browser cookies) nurtured in work under the umbrella of Internet Ethics. Notably, the Internet Architecture Board (IAB) created a policy in 1989 concerning Internet Ethics which offered general guidelines to prevent unethical Internet activity such as, gaining unauthorized access, wrongfully using computer-based information, and compromising the privacy of users (Internet Architecture Board,1989).

A fourth important area of academic scholarship on ethics and technology emerged from more recent multidisciplinary research and developments in nanotechnology, neurotechnology, and advances in military research. Beginning in the 1980s, seminal work like Drexler's (1986)*Engines of Creation: The Coming Era of Nanotechnology* helped raise public awareness about contributions and issues in nano-scale research and development. Nanotechnology and its applications have widely contributed in multiple disciplines (Computer Science, Engineering, Biology, and Chemistry) while raising ethical questions concerning health and safety issues and potential risks to the environment. In the context of military research, the engineering of biological and biotech weaponry have changed the landscape of military ethics. Hartle's (1989)*Moral Issues in Military Decision Making* was an important text that helped focus scholarly attention on new technological advances in the military and how it redefined relationships. Hartle's work provided a defining example of work on ethical aspects of military decision making at a time where technologies have become increasingly dominant in military action. Concurrently, the advent of neurological and cognitive research, artificial intelligence (AI) research, robotics, and systems research have raised new ethical questions about human cognition, artificial autonomous agency, moral rights and responsibilities of human creations, and technological control in society. nurtured scholarly work on social and ethical aspects of technology as seen in selected work in the Philosophy of Technology and technocritical scholarship.

By drawing on historical work underlying Technoethics, two complementary driving forces can be discerned, namely, a concerted effort to advance an ethical restructuring of technology and an effort to contribute to the technological enrichment of ethics. The result of these driving forces to identify and prioritize ethical aspects of new technological developments has the potential to leverage how individuals live and work within an evolving knowledge society dependent on technological progress.

The Rationale Underlying Technoethics

The rationale underlying concerted efforts to cultivate the field of Technoethics stem from multiple sources. First, the rise of potent technologies in diverse areas of human work and life have created new ethical dilemmas in society and various programs of ethical inquiry found in niche areas within multiple disciplines and fields from Engineering to Philosophy to the Social Sciences to Medicine. In the past, attempts to raise public understanding about the social and ethical aspects of new technologies have fallen short. This is partly due to a pervasive tendency within academia to create silos of information within fields and disciplines disconnected with one another. As a result, many opportunities to exchange ideas, leverage understanding, and create new knowledge have been missed. Technoethics overcomes this limitation because it is an interdisciplinary field of study and research with a relational orientation to technology and human activity. It attempts to bridge the gap in understanding of technological growth and those affected by it by bringing together all technology focused areas of ethical inquiry. In doing so, it provides a more comprehensive knowledge base reflective of the multidisciplinary framework that contributes to all areas of technoethical inquiry.

Second, Technoethics has a techno- and bio-centric (biotechno-centric) field of study and research. As such, it has a relational orientation to both technology and human activity allowing it to bring together diverse areas of ethical inquiry across all areas of human activity involving technology under a unified framework. This is to be differentiated from Applied Ethics where bio-centric ethical principles place living entities (biological) at the centre of ethical concern, thus limiting the scope of inquiry within technological contexts (Floridi, 1999). The advantages of placing both technology and biology as central are twofold. First, it assigns technology a new status and importance previously reserved for only living entities. Thus, Technoethics aligns with new areas of scholarly inquiry concerning artificial morality and current debates concerning the assignment of rights and responsibilities from designers and developers of technology as well as those affected by it. Second, it treats technology and biology as a general central organizing construct for dealing with all existing (and potential) technologies in relation to human activity and other living entities, rather than equating Technoethics with specific technologies. This follows a line of reasoning contributed by important scholars who recognized that ethical inquiry focused on specific technologies, like Computer Ethics, may disappear in the near future when computing becomes a mature technology and computer use becomes part of ordinary human action. As stated by Johnson (1999):

What was for a time an issue of Computer Ethics becomes simply an ethical issue. Copying software becomes simply an issue of intellectual property. Selling software involves certain legal and moral liabilities. Computer professionals understand they have responsibilities. Online privacy violations are simply privacy violations. So as we come to presume computer technology as part of the world we live in, Computer Ethics as such is likely to disappear (p. 45).

The field of Technoethics advances through concerted study and research on ethical issues surrounding all technology affecting human activity in the past, present, and future. Ethical issues are resolved (or redirected as ethical standards) as controversial new technologies become mature technologies integrated in society or rejected.

The justification for Technoethics derives from active engagement among a growing network of scholars working in diverse niche areas in technoethical inquiry, and concerted efforts to advance the best possible opportunities for scholars and researchers to share work on the unique social and ethical aspects of human activity affected by technology. "Technology raises special ethical issues, hence Technoethics deserves special status," as noted by Gearhart (2008). This distinguishes Technoethics from other approaches to ethical inquiry while building on work derived from key niche areas of multiple disciplines and fields concerned with ethical aspects of technology. As such, Technoethics represents a promising field of study and research to serve as the main organizing framework for past, present, and future scholarship in technology and ethics.

Defining the Field of Technoethics

In terms of defining Technoethics as a contemporary field of interdisciplinary research, Galvan (2001) defined it as the "sum total of ideas that bring into evidence a system of ethical reference that justifies that profound dimension of technology as a central element in the attainment of a "finalized" perfection of man." This definition expanded the scope of Technoethics into all fields and areas of human activity affected by technology. In a slightly different vein, Bao and Xiang (2006) defined Technoethics as the ethical basis for the global community. Building on these definitions and others, this book follows the *Handbook of Research on Technoethics*:

Technoethics is defined as an interdisciplinary field concerned with all ethical aspects of technology within a society shaped by technology. It deals with human processes and practices connected to technology which are becoming embedded within social, political, and moral spheres of life. It also examines social policies and interventions occurring in response to issues generated by technology development and use. This includes critical debates on the responsible use of technology for advancing human interests in society. To this end, it attempts to provide conceptual grounding to clarify the role of technology in relation to those affected by it and to help guide ethical problem-solving and decision making in areas of activity that rely on technology (Luppicini, 2008).

The general aim of Technoethics as a field is to advance technological relationships of humans with a focus on ethical implications for human life, social norms and values, education, work, politics, law, and ecological impact. In terms of theoretical framing and methodological tools, technoethical inquiry, as it is developed in this book (see Chapter 4), is a social systems theory and methodology for guiding technology assessment (TA) and technology design (TD). The conceptual work in this section places ethical questions related to technology on new ground. This, however, remains abstract without a discussion of the origins and key developments in technoethical inquiry.

DISCUSSION QUESTIONS

- What are ethical theories inadequate to deal with ethical problems created by the advent of new technologies?
- What does the philosophy of technology contribute to our understanding of technology and ethics in society?
- What can theories of technology and technocriticsl studies contribute to our understanding of technology and ethics in society?
- Which key historical developments can be considered precursors to contemporary Technoethics?
- The rationale underlying the field of Technoethics stem from what sources?

DISCUSSION

This chapter attempted to frame the emerging field of Technoethics within a philosophical and historical context. The first section sketched out the rise of Applied Ethics which brought philosophical inquiry into the realm of real world human practices and the pursuit of science and technology. It also reviewed pivotal work from the Philosophy of Technology and technocritical writings to demonstrate how scholars were became more aware of the need to connect technological development to human social and ethical entrenched in how people live, work, and play. This placed ethical questions related to technology on new ground requiring special status, laying the groundwork for the emergence of Technoethics. This also set the stage for core branches of Technoethics to emerge from sub-areas of Philosophy, the Social Sciences, and interdisciplinary domains of academic scholarship on ethics and technology.

REFERENCES

Achterhuis, H. (2001). *American philosophy of technology, the empirical turn.* Bloomington, IN: Indiana University Press.

Bao, Z., & Xiang, K. (2006). Digitalization and Global Ethics. *Ethics and Information Technology, 8*, 41–47. doi:10.1007/s10676-006-9101-7

Borgmann, A. (1984). *Technology and the character of contemporary life: A philosophical inquiry.* Chicago: University of Chicago Press.

Chadwick, R. F. (1997). *Encyclopedia of applied ethics.* London:Academic Press.

Drexler, E. (1986). *Engines of creation: The coming era of nanotechnology.* New York: Anchor Press/Doubleday.

Ellul, J. (1964). *The technological society.* New York: Knopf.

Feenberg, A. (1999). *Critical theory of technology.* New York: Oxford University Press.

Ferré, F. (1988). *Philosophy of technology.* Englewood Cliffs NJ:Prentice-Hall.

Floridi, L. (1999). Information ethics: On the theoretical foundations of computer ethics. *Ethics and Information Technology, 1*(1), 37–56. doi:10.1023/A:1010018611096

Floridi, L., & Sanders, J. (2003). Computer ethics: Mapping the foundationalist debate. *Ethics and Information Technology, 4*(1), 1–24. doi:10.1023/A:1015209807065

Foot, P. (1978). *Virtues and vices*. Oxford: Blackwell.

Fuller, S. (2005). *The philosophy of STS*. New York: Routledge.

Galván, J. (2001). *Technoethics: acceptability and social integration of artificial creatures*. Retrieved June 30, 2007 from http://www.eticaepolitica.net/tecnoetica/jmg_acceptability%5Bit%5D.htm

Gearhart, D. (2008). Technoethics in Education for the Twenty-First Century. In R. Luppicini & R. Adell, (eds.), *Handbook of Research on Technoethics*, (pp.263-277). Hershey: Idea Group Publishing.

Hartle, A. (1989). *Moral issues in military decision making*. Lawrence, KS: University of Kansas Press.

Heidegger, M. (1977). The question concerning technology. In W. Lovitt, (Ed.), *The question concerning technology and other essays* (pp.13-39). New York: Harper and Row.

Higgs, E., Light, A., & Strong, D. (2000). *Technology and the good life?* Chicago: University of Chicago Press.

Hobbes, T. (1651). *Leviathan, with selected variants from the Latin edition*. Indianapolis: Hackett publishing.

Ihde, D. (1979). *Technics and praxis*. Dordrecht: Reidel.

Internet Architecture Board. (1989). Retrieved June 4, 2007, from http://tools.ietf.org/html/rfc1087

Jacobs, J. (1961). *The death and life of great american cities*. New York: Random House and Vintage Books.

Johnson, D. (1985). *Computer ethics*. Upper Saddle River, NJ: Prentice-Hall

Johnson, D. (1985) *Computer ethics*. Upper Saddle River, NJ: Prentice-Hall.

Johnson, D. G. (1999). Computer ethics in the 21st century. In [Rome, Italy.]. *Proceedings of, ETHICOMP99*, 44–60.

Jonas, H. (1985). *On technology, medicine and ethics*. Chicago: Chicago University Press.

Kant, I. (1785). *Grounding for the metaphysics of morals* (3rd Ed.). Indianapolis: Hackett.

Kaplan, D. (Ed.). (2004). *Readings in the philosophy of technology.* New York: Rowman & Littlefield.

Kuhn, T. (1962). *The structure of scientific revolutions.* Chicago: University of Chicago Press.

Latour, B. (2005). *Reassembling the social: An introduction to actor-network-theory.* Oxford: Oxford University Press.

Latour, B., & Woolger, S. (1979). *Laboratory life, the social construction of scientific facts.* Los Angeles: Princeton paperbacks.

Locke, J. (1690). *An essay concerning human understanding.* England: Oxford University Press.

Luppicini, R. (2008). Introducing technoethics. In R. Luppicini & R. Adell (eds.), *Handbook of research on technoethics (pp. 1-18).* Hershey: Idea Group Publishing.

MacIntyre, A. (1985). *After virtue.* London, Duckworth: Jacksonville University.

McGinn, R. (1978). What is technology. *Research in Philosophy and Technology, 1,* 179–197.

Mitcham, C. (1994). *Thinking through technology.* Chicago: University of Chicago Press.

Nozick, R. (1974). Anar*chy, state, and utopia.* New York: Basic Books.

Oxford Concise Dictionary. (2006). *Ethics.* Oxford, UK: Oxford University Press.

Passmore, J. (1974). *Man's responsibility for nature.* London: Duckworth.

Pinch, T. J., & Bijker, W. E. (1987). *The social construction of technological systems: New directions in the sociology and history of technology.* Cambridge, MA: MIT Press.

Rapp, F. (1974). *Contributions to a philosophy of technology, the case of preparative chemistry.* Germany: University of karlsrue.

Rawls, J. (1970). *A theory of justice.* Cambridge: Belknap Press of Harvard University Press.

Rolston, H. (1975). Is there an ecological ethic? *Ethics, 85,* 93–109. doi:10.1086/291944

Singer, P. (1986). *Applied ethics.* New York: Oxford University Press.

Stehr, N. (2002) *Knowledge & economic conduct.* Toronto: University of Toronto Press.

Stone, C. D. (1972). Should trees have standing? *Southern California Law Review, 45,* 450–501.

Tavani, H. T. (2007). *Ethics and technology: Ethical issues in an age of information and communication technology.* Hoboken, NJ: John Wiley & Sons.

Walzer, R. (1977). *Just and unjust wars: a moral argument with historical illustration.* Cambridge, MA: MIT Press.

Weizenbaum, J. (1976). *Computer power and human reason: From judgment to calculation.* New York: Freeman.

Wiener, N. (1948). *Cybernetics: Or control and communication in the animal and the machine.* Cambridge, MA: The Technology Press.

Wiener, N. (1954). *Human use of human beings.* New York: Houghton Mifflin.

Winner, L. (1977). *Autonomous technology: technics-out-of-control as a theme in political theory.* Cambridge, MA: MIT Press.

ADDITIONAL READING

Borgmann, A. (1999). *Holding onto reality.* Chicago: University of Chicago Press.

Broad, C. D. (2000). *Ethics and the history of philosophy.* UK: Routledge

Bunge, M. (1967). *Scientific Research II: The Search for Truth.* New York: Springer.

Bunge, M. (1976). The philosophical richness of technology. *Proceedings of the Biennial Meeting of the Philosophy of Science Association. Volume 2: Symposia and Invited Papers (pp. 153–172*

Jan Van Dijk. (2006). *The network society.* London: Sage. Second Edition.

Jonas, H. (1984). *The imperative of responsibility, in search of an ethics for the technological age.* Chicago: University of Chicago Press.

Latour, B. (1996). *Aramis, or the love of technology.* Harvard University Press, Cambridge Mass., USA.

O'Riley, P. (2003). *Technology, culture, and socioeconomics: A rhizoanalysis of educational discourses.* New York: Peter Lang

Rosen, F. (2003). Classical Utilitarianism from Hume to Mill. Routledge.

Stankosky, M. (2004). *Creating the discipline of knowledge management: The latest in university research.* Butterworth-Heinemann,Touraine, A. (1988). *Return of the actor.* Minneapolis. University of Minnesota Press.

Section 2
Technoethics Deployment Framework in the Knowledge Society

Chapter 3
Technological Consciousness and Moral Agency

INTRODUCTION

As in manufacture so in science–retooling is an extravagance to be reserved for the occasion that demands it. The significance of crises is the indication they provide that an occasion for retooling has arrived. –Kuhn, 1962, p.76.

Is it possible to explain social and ethical aspects of technology in society without considering the human minds and actions intertwined within technological advances? Can legal and ethical questions concerning agency in autonomous machines be addressed without meditating on the conditions of consciousness required for agency? The answer to both these questions is no. A persistent problem

DOI: 10.4018/978-1-60566-952-6.ch003

in the study of technology today is the lack of attention to the nature of the human mind and how it fits into the real world of technology. Scholars have tended to draw on philosophical-sociological theory and group themselves into camps (e.g., technological determinism, social constructivism, actor-network theory, etc.). Most of these theories, however, fail to address the human side of technology that lies within 'individual' minds and bodies that affect and are affected by technology at a deeply personal level. In other words, the mental life of human subjects is not a core consideration in the study of technology in society. What remains is a persisting problem within a continually advancing technological society to understand the relationship between technology, consciousness, and society.

Technology is not only important in contemporary society, it is also at the root of what it means to be human. As stated by David Nye (2007), "Technology matters because it is inseparable from being human. Devices and machines are not things "out there" that invade life. We are intimate with them from birth, as were our ancestors for hundreds for generations" (ix). Technology is core to human development and a key focus for understanding human life, society, and human consciousness. The longstanding importance (and challenge) of technology to human life is attested by its pervasive entrenchment in human consciousness, life and society throughout history.

Now, more than ever before, there is a crucial need to consider how the human mind connects with technology and society in order to shed light on the complex relation between humans, technology, and society. There is also a need to better ground technological theories within technological processes wherever they occur within the mind, body, and world. This can be achieved by framing the study of technology within a relational stance to more tightly ground the study of technology and society in human life and visa versa. This chapter delves into the characteristics of technological consciousness rooted within a knowledge-based society fuelled by advancing science and technology and the need for a technoethical framework for leveraging understanding and guiding societal practices. It highlights how technology is a core source of meaning in human life and a driving force at the root of autonomous agency in humans and autonomous machines.

BACKGROUND

Technology Conceptualized

As with many concepts so deeply connected with human values and interests, "technology" is a term that has been defined in many ways and worn smooth by a thousand tongues from different historical periods and social contexts. What is particularly interesting is how our understanding of technology (under various labels) has changed throughout history, just as our understanding of ourselves and the world we live in has changed. One way to discern the meaning of technology to human life and society is to look at how it has been conceptualized at various points in human history under various labels (e.g., techné, technik, technology).

In terms of historical framing, there are four main historical periods where conceptualizations of "technology" contributed to the entrenchment of technology in the minds and actions of humans. Each of these conceptualizations is sketched out below with key elements described.

First, early Greek civilization defined technology as techné. Techné refers to "the set of principles, or rational method, involved in the production of an object or the accomplishment of an end; the knowledge of such principles or method; art" (Stanford Encyclopedia of Philosophy, 2008). Within the early Greek tradition equated techné with the Mechanical Arts (e.g., blacksmithing, navigation, agriculture, hunting, medicine). This had both negative and positive connotations since formal knowledge of truths and virtues were valued more highly in the work of Socrates and the Platonists. This grounded the concept of technology in applied knowledge, skills and procedures carried out by individuals in core areas of work and life. Techné was valued more highly within the Roman Empire and was equated with the human power to construct homes and roads. Cicero believed was the creation of a second nature (*Alteram Naturam*) as reflected in the following statement. "We sow corn and plant trees. We fertilize the soil by irrigation. We dam the rivers, to guide them where we will. One may say that we seek with our human hands to create a second nature in the natural world" (Cicero, 1972). This emphasis of techné as 'second nature' is significant because it connected social development and technological activity in the world with the internal workings of human nature. In other words, it placed technology at the centre of human meaning and social progress.

Second, the modern conceptualization of technology as invention materialized in the 17th century in Francis Bacon's futuristic vision of a perfect society governed by engineers and scientists in the Saloman's House, what Bacon viewed as, "the noblest foundation (as we think) that ever was upon the earth; and the lanthorn of this kingdom" (Bacon, 2008). In this imaginary house, new inventions (e.g., medical procedures, loudspeaker systems, submarines) represented the engines of progress with the potential to transform society into the perfect society where material needs were satisfied and poverty did not exist. The actualization of this view of technology as invention became entrenched in modern life as early as 1662 with the founding of the Royal Society in London, dedicated to leveraging the role of science and invention in society (Nye, 2007). Although limited to technology as invention, this modern view helped raise the importance of technology in society and pave the way for further conceptualizations of the term more closely aligned with science and innovation in society.

Third, in the late 19th and early 20th century, the German term "technik" (English trans. Technic) was used extensively by engineers in their professional work to describe the totality of processes, tools, machines, and systems employed in the practical arts and Engineering. This term was further popularized in the seminal the writings of Max Weber with its meaning broadened even further (beyond the practical arts and engineering) by Lewis Mumford who equated technology with all systems of techniques and machines underlying a civilization. Mumford's (1934), *Technics and Civilization* viewed technology (technological complex) as an indicator of social and historical evolution defined by periods of production: Eotechnic (before 1750) was dominated by water and wood industries, paleotechnic (1750-1890) was dominated by coal and iron industries, and neotechnic (1890) was dominated by electrical and alloy industries. This view of technology (as technics) helped to broaden understanding of technology and place it at the center of social life in close connection to social progress and societal change. Mumford indicated, "The machine cannot be divorced from its larger social pattern, for it is this pattern that gives it meaning and purpose" (p. 100).

Forth, from the mid 20th century to present time, the term "technology" has become unclear due largely to technological critique and a sudden narrowing of its meaning. Rapid advances in technology provoked a negative reaction from scholars who viewed technology as a controlling force in society with the potential to destroy how people live (technological determinism). Heidegger (1977) warned people that technology was dangerous in that it exerted control over people through its mediating effects, thus limiting authenticity of experience in the world that defines life and gives life meaning. Beginning in the 1950s, technology was defined in terms of complex systems and machines by professionals connected to technological innovation. This narrowing of meaning continued in the 1990s when technology became equated with electronic devices (e.g., computers, radios, telephones, etc.). The result of this conceptual narrowing has left the meaning of technology unclear and obscure.

In drawing from the above historical conceptualizations of technology, the German derived notion of technics builds on previous conceptualizations of technology and offers the broadest conceptionalization for framing social and ethical studies of technology. This is because it captures the breadth of individual and social meaning required to appreciate technology in society. It is noteworthy to mention that the German derived notion occurred at a time period which preceded the advent of new electronic technologies and progress in Psychology and Cognitive Sciences. This work emphasized the importance of studying internal human operations and patterns using scientific methods. In light of new developments Mumford's view of technology (as technic) can be extended to include new technological innovations and scientific progress in human psychology--Technology cannot be divorced from its individual and social patterns that gives it meaning and purpose. This view of technology embraces the totality of factors that currently define human meaning and social progress, especially within countries like the United States that highlight individual rights. At the same time, it entrenches technology in the roots of the human mind and history. This discussion helps illustrate that the conceptualization of technology is not a new phenomenon within a rapidly advancing technological society. Rather, it is an intimate part of the human condition, deeply entrenched in all human history, society and mind.

The Cognitive Turn in Technology Studies

The cognitive study of technology is rooted in cognitive sciences and the philosophy of science and technology that began taking shape in the early 1960s. At this time, a new awareness of human influence on the development of science and technology was beginning to emerge. This was exemplified by Kuhn's (1962) *The Structure of Scientific Revolutions*. Kuhn helped draw attention to the importance of scientific reasoning and human imagination (invention) in the advancement of science, "Philosophers of science have repeatedly demonstrated that more than one theoretical construction can always be placed upon a given collection of data. History of science indicates that, particularly in the early developmental stages of a new paradigm, it is not even very difficult to invent such alternates" (p. 76). Kuhn and subsequent scholars resituated conceptions of science and technology as human enterprises governed by human decisions and thinking. Some scholars in this period adapted extreme relativistic stances rejecting claims to generalized methods in the pursuit of science such as Paul Feyerabend (Feyerabend, 1975, 1978). Subsequent post-1960s scholarship in the philosophy of science and technology helped transform previous conceptions of science as governed solely by rationality and objectivity while opening up new possibility in the study of science, technology, and mind.

The coming of cognitive sciences in the 1960s can be juxtaposed with the changing nature of philosophy of science and technology scholarship. Beginning in the 1960s, cognitive science began emerging when scholars from multiple disciplines started advancing theories of the mind and intelligence grounded in complex representations and computational procedures. Following the trend in the philosophy of science and technology, early work in cognitive science tended to focus on processes of scientific thinking and discovery. Hebert Simon, an early proponent of artificial intelligence research, began investigating processes involved in scientific discovery in an effort to discern the cognitive nature of scientific thinking (Simon, 1966). This inspired subsequent scholarship aimed at understanding the cognitive roots of technoscientific reasoning exemplified in works such as Giere's (1988)*Explaining Science: A Cognitive Approach* and Michael Gorman's (1992)*Simulating Science: Heuristics, Mental Models, and Technoscientific Thinking.* The founding of the Cognitive Science Society and the creation of the journal Cognitive Science in the mid-1970s helped spur additional academic interest in scientific studies of mental life and technoscientific thinking as found within a variety of works including, Churchland's (1989) *Neurocomputational Perspective: The Nature of Mind and the Structure of Science*, McClelland, Rumelhart, and the PDP Research Group's (1986) *Parallel Distributed Processing: Explorations in the Microstructure of Cognition*, Clark's (2001) *Mindware: An Introduction to the Philosophy of Cognitive science*, Dreyfus' (1992) *What Computers Still Can't Do*, and Johnson-Laird's (1988) *The Computer and the Mind: An Introduction to Cognitive Science.*

Amassing scholarship within philosophy of science and technology and cognitive sciences from the 1960s onward has provided many new opportunities for advancing the study of science and technology that did not exist 40 years ago. This is particularly important when examining broader social and ethical considerations. First, cognitive studies allowed more sophisticated ways and helpful concepts (collective intelligence, distributed cognition) to examine questions of human agency and ethical responsibility. Second, cognitive approaches to studying science and technology are complimentary to sociological theories of technology. Third, cognitive based accounts place technological thinking at the centre of mental and social life. The sophistication, theoretical compatibility, and technological focus has helped ground cognitive approaches to science and technology studies as exemplified in Ascott's (2003)*Telematic Embrace: Visionary Theories of Art, Technology and Consciousness*, Ranade's (1998)*The Technology of Consciousness*, and Gorayska and Mey's (2004)*Cognition and Technology: Co-existence, Convergence, and Co-evolution.* This work is part of expanding body of scholarship that places technology at the core of mental life and social action. The next section attempts to remove some of the persisting stigma that has blocked the development of strong connections between technology and consciousness.

COMMON MISUNDERSTANDINGS ABOUT CONSCIOUSNESS AND TECHNOLOGY

Common Misunderstandings

Common misunderstandings often surround important concepts like "consciousness" and "technology", which have been defined in so many ways by different scholars over the years. The resulting lack of conceptual clarity about the relation between consciousness and technology preventing the building of meaningful inroads between these two concepts. This book is based on the assumption that a better understanding of technology and its connection to mind and society is needed to properly frame social and ethical studies of technology. This section delves into pervasive misunderstandings and conceptual narrowing has left the meaning of consciousness and technology unclear.

Probing into the nature of consciousness (often articulated as the mind-body problem)has been a preoccupation with philosophers and scientists around the world for decades. One view is that mental states are not states at all but are autonomous agents (Minsky, 1985) within a networked consciousness. Another view is that mental processes exist as a series of mental drafts of consciousness which bubble to the surface (Dennett, 1991). What is consciousness and where do we find it?

On the surface, consciousness appears simple and self evident. When individuals reflect inward (introspect) there is a rich inner life of subjective experience— memories, moods, concepts, hopes, fears, colors, tastes, likes, dislikes, etc. All of this is commonly believed to reside somewhere in the head amongst the neurological processes and brain structures. In reality, these common sense views are among a number of misconceptions that have blocked progress in efforts to understand consciousness and how it shapes and is shaped by technology in society. Clarification of these misconceptions about the nature of consciousness and technology is needed to allow a clearer picture of consciousness and technology that overcomes a schism that persists in the academic literature. The main misconceptions are listed as follows:

- Consciousness is only in the head
- Technology is not part of consciousness
- Technology controls society and consciousness
- Society controls technology and consciousness

Consciousness is Only in the Head

When asked where consciousness resides, many people will quickly motion to their head and say, "It happens here!" This is false. Once more, individuals also assume it is the same in others. When speaking with someone and looking into their eyes, there is an assumed space somewhere behind the eyes that is addressed. While someone is conscious, there are neurological processes operating inside one's head but this is not consciousness and does not dictate where consciousness is occurring? If one were racing their BMW, internal neurological patterns are operating inside one's head but the conscious experience is not. If someone imagines they are racing their BMW, the same neurological processes are operating in the head but the

BMW and the conscious experience of racing are not to be found in the head, but rather in the BMW racing around the streets. Consciousness is responsible for the creation of new conscious relations wherever imagined, be it in the head, on the street, or in the past.

Technology is Not Part of Consciousness

Technology in the mid-20th century was defined in terms of complex systems, machines, and electronic devices providing the vision of technology as objects in the external world and outside human beings. How could a computer or a telephone be considered in any way connected to human consciousness or mental life? Could a theoretical model created in the mind of an engineer be connected to technology? The strawman response using this narrow definition of technology is no. But, as discussed in the previous section, the conceptualization of technology has gone through drastic changes, the most recent conceptualization being the most narrow and not the most useful.

Technology Controls the Human Mind and Society

On the other extreme end of the technology theory spectrum is the view that technology determines thoughts and actions. Heidegger (1977) warned the world that "we do not yet hear, we whose hearing through and seeing are perishing through radio and film under the rule of technology" (p. 48). During the 20th century advances in production and electronic technology had a massive impact on society which provoked a negative reaction from scholars and the rise of technological determinism in academic scholarship. Part of the problem of this movement that is now dying out is that it mainly focused on technology in its narrowest conceptualization as external devices and tools.

Another facet of the problem with technological determinism is the current softer argument of technological control of human experience. This new aspect of the technological control debate is particularly prevalent in Clinical Psychology where Internet addiction, cell phone addiction, and dissociative identity disorder are creating new profits for practicing psychologists. Part of the problem is that this trend capitalizes on pervasive public misconceptions concerning technological determinism, giving rise to widespread fear among individuals with real mental issues to deal with. It is not that the problems identified are not real, but they are mislabelled. Mental traumas linked to technology are not to be controlled by treating the symptom while ignoring the sickness. Such problems cannot be resolved by removing devices from ones' environment, but by recognizing that technology is rooted in consciousness as an integral part of mental life for everyone. This understanding will most likely alter how both patients and psychologists deal with the trials and tribunes of living with technology.

Society Controls Technology and Consciousness

Social constructivist accounts typically argue that technological determinism fails to take into consideration the social shaping of technology responsible for guiding which technologies get selected and developed. The role of human agency in technological decision making began to be recognized through social constructivist perspectives. This, however, draws attention to a further problem that neither the deterministic or social constructivist perspectives address. What is not typically highlighted is the fact that the social construction of technological decisions arises from individuals' technological processes (arguments, constructions, etc.) involved in various networks of social action and normative frameworks. The social constructions themselves presuppose the existence of a technological consciousness within individuals capable of social construction. Part of the problem of the deterministic and social constructivist accounts is that they fail to acknowledge the complex relational nature of technology as an operation within mind and society. This realization shifts the focus on technology to its origins within the human mind as explained through the theory of technological consciousness.

TECHNOLOGICAL CONSCIOUSNESS

Technological Consciousness

Debates concerning all aspects of the mind, mental processes and their organization are beyond the scope of this chapter. But, what most current theories of consciousness agree on is that there are conscious states or phenomenon that give rise to potentially observable acts that can be reproduced. Rather than focusing on all aspects of the human mind and its operation, a focused study of consciousness within technology studies is primarily concerned with those technological processes originating in the mind that can lead to the proliferation of technologies powerful enough to alter nature and our place in it. It begins with a conceptualization of technology suitable for addressing the broad and changing needs of contemporary society.

Broader conceptualizations of technology (as technic) come closer to approximating a more useful conceptualization of technology as it exists in life and contemporary society. This provides a view of technology as a set of relational processes which originate in the mind as operations of consciousness (henceforth technological consciousness) which manifest themselves within individual and social activity. It is easy to imagine an engineer thinking through a new machine design that improve on an existing model, sketching it out on a pad or computer screen to share with other professionals, followed by the construction of a prototype based on the new design specs. Through testing of the machine prototype, problems and possible modifications can be realized, communicated to others, and incorporated in the construction of the next prototype. These are all technological processes which a common root within technological consciousness. In this sense, humans cannot be separated from technology because it is an inherent part of consciousness and meaning in life. This is true regardless of whether the technological processes are imaged in the mind or manifested in the world.

It must be noted that technological consciousness is not a complete picture of the human mind. There are many aspects of the human mind that are conscious but not connected to technological consciousness, such as moods and memories. One can be conscious of being in a certain mood or state without creating new relations that change consciousness. Similarly, one can remember familiar experiences from the past without any conscious effort to create new relations with new elements in consciousness. These aspects of the human mind are important aspects of the human condition but not directly connected to technological consciousness because they do not intend towards the creation or invention of anything new. This discussion focuses mainly on technological consciousness because it is considered to be central to human innovation and at the core of social and ethical aspects of technology in life and society.

What is technological consciousness and how does it work? Technological consciousness consists of mental operations which generate relations between familiar and unfamiliar elements to create new knowledge in the form of technological processes. The main operation of technological consciousness is to generate new understandings required as human experiences and society become more complex. In other words, Technological consciousness is at the seat of evolving consciousness in a rapidly changing world of individual experience, social configurations, and environmental conditions. The basic structure of technological consciousness can be defined as: evolutionary, relational, situated, organizational, aspectual, and integrative.

Technological Consciousness is a Relational Operation

Technological consciousness generates technological processes including actions, speech acts, symbols, and language. These technological processes are rooted deep within our minds, our social contexts, and our history. It is through technological consciousness and the acts of technological processes that languages grow and theoretical models are created. The invention of the phonetic alphabet represents one of the most powerful technological processes, but not the first. Dating back to the Paleolithic Period and the coming of Homo sapiens approximately 200,000 years ago, early man successfully sharpened the edges of stones and attached these stones to sticks for digging, hunting, and combat. The relational pattern of this configuration of objects was created to stand for a new grouping of familiar objects (sticks and stones). The art of spear-making evolved through continual observation and improvement on existing designs observed, that is, by creating a theory between an existing model of a spear and what it represents. The same can be said of most if not all crafts, which embodies the early action oriented conceptualization of technology as "teche" envisioned by the early Greeks. Oral tribal cultures created sounds to represent objects, states, and events which could be communicated to others and passed on from person to person and from generation to generation. To continue with the same example, an understanding of a spear would be generated by substituting a sound for the familiar object (spear). In this sense, technological consciousness is the building block for the creation and communication of new tools, language, and theory. This is the conceptual realm of technology that humans experience supported by broad conceptualizations of technology (i.e., techné, technic) discussed earlier.

Technological consciousness is a relational operation rooted in the human mind, which gives rise to mental and physical acts (technological processes) that are observable (or potentially observable) and which can be reproduced either by the individual or another. The operation of technological consciousness can be likened to the lighting of the Olympic torch. When the torch is lit by one person, that person can run with it and it can be seen by that person and others as the runner passes by. The torch is observed by onlookers and can be passed off to another. This captures the power of technological consciousness to create and re-create inventions which in turn, can shape and be shaped in society.

Technological Consciousness is Organized

Technological consciousness consists of organized sequences of relational operations, rather then separate and disjointed pieces of conscious experience. For instance, in evaluating a new instrument used in medical examinations, an evaluator does not have a conscious experience of only the technician's voice, a separate experience of sensation of the coldness of the instrument, and a separate visual experience of what is hanging on the office wall. Instead, the evaluator is simultaneously conscious of an organized unity of experience with all its diverse features. At the same time, one is conscious of turning on the instrument being used, adjusting the settings, and performing an instrument test run on a patient. This requires that conscious experiences are also organized as a temporal sequence with a beginning, a middle, and an end. Kant (1787) referred to this quality of technological consciousness as the transcendental unity of apperception. Without these features of technological consciousness (and consciousness in general) it would be difficult to make sense out of experience. The organizational quality of technological consciousness ensures a continuity of imagined experiences and how each element interrelates in space and time. This also helps further demonstrate that even at the most basic level of conscious experience, the relational operations of the human mind are present.

Technological Consciousness is Situated

Human beings are often frustrated when trying to remember the name of a forgotten acquaintance or what one's first grade teacher looked like. It often helps one remember when a face and name can be matched together or when a place can be matched with a person. The reason is that consciousness is situated within the location one imagines. In this way, technological consciousness possess the quality of spatialization With each new situation experienced, the relational operation of consciousness attributes a location in time and space. That is why remembering a partial aspect of a conscious experience (like remembering a place where an acquaintance was met) can help one piece together all aspects of that experience (like the acquaintance's name or other attributes). A great deal of work in Cognitive Psychology has been invested in figuring out the underlying mechanisms affecting memory and consciousness.

The situatedness of consciousness is particularly important because many of the technological processes created through the relational operations of the human mind occur in social organizations and the environment. This allows technological processes to act in social institutions and alter the natural environment while remaining rooted in human minds. For instance, when a computer designer creates a new interface for synchronous communication, users situate the conscious experience in virtual space and attribute qualities to another user context beyond the realm of one's own consciousness and the computer screen. The situatedness of technological consciousness allows individual computer users to create a context for situating conscious experience that did not exist before the advent of computers. The same was the case for the telephone. From this, one can see that technological consciousness is responsible for creating the underlying contexts as well as the contents of conscious experience.

Technological Consciousness is Aspectual

Individuals are never conscious of everything within an experience. When walking into a crowded bar, one may remember the single red-head sitting at the bar among dozens of blonds. Or maybe, one familiar face on the crowded dance floor sticks out in one's mind. People often pay attention to particular aspects, ones that are salient, or familiar, or of interest. In other words, technological consciousness is aspectual. In working with a multidisciplinary research team in a government lab, an engineer may be aware of instrument readings and levels of performance, a computer designer my be aware of graphics and user interface elements, and a project manager may be aware of team interactions and research costs. Individuals are not conscious of things in themselves, only aspects of things. Because this is the case, technology cannot be limited to consciousness but must manifest itself in technological processes (e.g. language, acts, constructions) that have the potential to be shared with others. This is especially true in a technologically advanced society where large scale innovation requires many individuals with specialized expertise. Specialists select from a collection of possible aspects of an event, thing or behaviour, which comprises their knowledge of it. It is in the coming together of different relational operations of technological consciousness through technological processes that humans are able to exchange their partial perspectives, and sometimes create innovative new tools and techniques that influence society.

Technological Consciousness is Integrative

Technological consciousness has many integrative features by which relational operations create new technological processes in the mind and world. Key integrative features are discussed here: assimilation, conversation, substitution.

Assimiliation. Assimilation is perhaps the most basic feature of consciousness where a new conscious experience that is unfamiliar is integrated with familiar conscious experiences even though it is different. Assimilation is a relational operation of consciousness that allows individuals to create new conscious experiences based on what is already there. If a mechanic experienced in American motors has to fix a German made Volvo, then there is considerable base knowledge and conscious experience of one motor that will help expediate the repair if experience derived from working on the Volvo is integrated with previous conscious experiences with American motor repair.

Substitution. Within the technological consciousness, substitution is another key relational operation, one that is at the base of language creation, theory building, and the use of other symbols and metaphorical devices. Language is built up by relational operations of technological consciousness through vocabulary/sounds which stand for phenomenon in the social and physical world. Language is a formidable technological process in that it allows individuals to communicate complex conscious experiences using a simple codified set of terms. This allows complex experiences to be expressed and shared independent of the actual conscious experience of one individual. In this sense, language is a main technological process that helps ground technological consciousness of individuals in the social and natural world. A theory is a prescribed relation between a conceptualization and what that conceptualization is intended to represent. A theory is a complex form of substitution where many substitutions can be interlinked in an effort to describe something.

Conversation. Within the human mind, individuals imagine themselves as the main speakers in the conversation of their lives. This set of core relational operations within the human mind can be referred to as the conversational quality of mind. Like all conversations, the topic (conscious contents) continually changes as the conversation evolves with different mental processes (technological processes) entering this ongoing conversation. The conversational quality of mind allows new conscious experiences to be integrated into the conversation of ones life. Within a technological consciousness, it makes no sense to speak of the "I" except in relational terms where organized, situated, aspectual conscious experiences come together and are integrated into the conversation of mind. This integrative feature of technological consciousness is core in that it forms a meta-level of relational operations that guide technological processes that require definite commitments and actions of the individual. In this way, the conversational quality of the mind is at the core of technological consciousness. Without it, the coordination of technological processes between individuals in society that affect the real world (i.e., production, inventions, new techniques) could not occur and technological consciousness would be disconnected from the social world.

This discussion articulated a theory of technological consciousness focusing on the interplay of technological processes that transform minds, bodies and society. This is intended to ground the study of technology based on the conviction that technology is deeply entrenched in the evolution of consciousness, bodily experiences, relationships, social institutions and identity. Technological consciousness explores the role of technological processes in our mental lives and in our evolving ideas about minds, bodies, and reality. It is based on the assumption that technological consciousness is a relational operation rooted in the human mind.

HUMAN AND ARTIFICIAL AUTONOMOUS AGENTS

As is developed in the next chapter (chapter 3), technology can be treated as a social system of human-technological relations open to pragmatic inquiry. The conceptualization of technological consciousness as a relational operation rooted in the human mind also provides a conceptual base for studying autonomous agency in humans and artificial entities (i.e., cybernetic organisms, autonomous software, autonomous robots). This is because technological consciousness, as part of the system of technology, gives rise to technological processes which are re/producible in individual minds, activities, and artifacts. It is the technological processes reproducible in artifacts as the locus of inquiry that are most relevant to questions of autonomous agency.

Following pragmatic oriented systems methodology, the meaning of artifacts derives from the relations between humans and their artifacts. As explained by McDermott (1976)," Artifacts, then, are human versions of the world acting as transactional mediations, representing human endeavour in relational accordance with the resistance and possibility endemic to both nature and culture". Many of these artifacts (as human constructions) can be classified as autonomous agents. An agent (something or someone with the power to affect change to other agents and/ or the environment) is an autonomous agent to the extent it can sense and interact within a changing environment. Franklin and Graesser (1996) defined an autonomous agent as, " a system situated within and a part of an environment that senses that environment and acts on it, over time, in pursuit of its own agenda and so as to effect what it senses in the future". In other words, autonomous agents (human or artificial) are systems defined by their ability to act independently without supervision of a human. Technological conscious attributed to humans can be re/produced in other autonomous agents whether human or artificial.

There are a variety of artificial autonomous agents types (see Franklin and Graesser, 1996), but the one's of central importance in Technoethics and questions of moral agency are cybernetic organisms, autonomous software agents, and autonomous robots. Cybernetic organisms (cyborgs) are natural entities with technological augmentation. Although science fiction often portrays cyborgs as machine like entities with human characteristics (e.g., Cylons, Replicants, Terminator, etc.), cyborg refers to human beings with technological enhancements. Types of technological enhancements include, artificial hearts, cochlear implants, biosensors, prothstetic limbs, etc. Autonomous robots (not to be confused with remote controlled robots) are robots where the robot programming is in control of the robot's actions. Sullins (2008) describes two types of autonomous robots as likely candidates for autonomous moral agency, namely, mobile robots and affective robots:

Mobile robots can display autonomy, continuous monitoring of their environment, motility, modest levels of communication and flexible machine learning. Affective robots are like mobile robots except they attempt to simulate emotions and expressions in order to enhance communication with the human agents it interacts with (p.207).

Autonomous software agents are autonomous programs (i.e., viruses, task agents) that exist in virtual environments (I.e. computers, networks, WWW). They are usually designed to assist human users or cause damage. Because of the popularity and potency of increasing sophisticated artificial autonomous agents of various types, there is an expanding interest within Technoethics to assign moral responsibility to technology creators and creations.

TECHNEOTHICS AND THE ROLE OF ARTIFICIAL MORAL AGENCY

Although moral (and immoral) aspects of technology have been a major focus in popular science fiction writing and film (e.g., Blade Runner, 2001 Space Odyssey, I Robot, etc) for many years, scholarly efforts to assign moral values to technology is rooted in early Technoethics and the work of Mario Bunge who assigned moral agency to technologists (1977):

Because of the close relationships among the physical, biological and social aspects of any large-scale technological project, advanced large-scale technology cannot be one-sided, in the service of narrow interests, short-sighted, and beyond control: it is many-sided, socially oriented, farsighted, and morally bridled (p. 101).

Within this early work in Technoethics, Bunge (1977) focused on the technologist as technology creator with the locus of social and ethical responsibility focused mainly on the technologist. This early work was important because it highlighted ethical aspects of technology at the core of technology oriented professions. While this early work was important in grounding Technoethics and the moral agency of technologists, it did not fully address the moral agency within the artifacts themselves. Rather, artifacts were treated as morally inert.

Technoethics did not focus a great deal of attention on moral agency in both technologists and technology until recently (see Luppicini & Adell, 2008). This is partly due to the state of technology advancement which changed drastically from the 1970s to today. In particular, the rise of ubiquitous computing in the last decade has provided more flexible technology integrated in more areas of society with higher capacities for autonomic functionality (Weiser, 1999). Weiser (1999) posited that powerful computing technology has given rise to artificial agents which have been integrated into many aspects of life in a way that it is implicit and invisible to the user (i.e., smart homes). Weiser believes that these artificial agents should be closely monitored to ensure that they enhance the humanness of our world instead of merely focusing on efficiency. Other work attempts to redefine the criteria used to determine what constitutes moral agency in artificial agents. As stated by Stahl (2006):

An analysis of the concept of responsibility shows that it is a social construct of ascription which is only viable in certain social contexts and which serves particular social aims. If this is the main aspect of responsibility then the question whether computers can be responsible no longer hinges on the difficult problem of agency but on the possibly simpler question whether responsibility ascriptions to computers can fulfill social goals. (p. 205)

This suggests that key questions about moral agency in Technoethics are un-resolved with much work still to do on core questions of interest such as, What does it mean to be a moral agent and what standards should a moral agent follow? What moral value, rights and/or responsibilities can or should be attributed to an artificial moral agent. It is noteworthy that work in information ethics and the use of information theories are becoming a vital component of technoethical inquiry to be used as tools for examining ethical aspects of information and information systems to theorize about the morality of artificial agents (Floridi & Sanders 2004). This has helped highlight the importance and flexibility of information within human and artificial agents. At the same time, there is need to rely on other tools to avoid repeating past failed attempts at providing computation theories of autonomous agents which reduce human consciousness to a machine language and reduce the human condition to a functional equivalence of autonomous machines. Much of the current work under information ethics and the reliance on information metaphors must be juxtaposed with other approaches in Technoethics to guide technoethical inquiry (see chapter 4).

DISCUSSION QUESTIONS

- What are some of the common misunderstandings about consciousness and technology and how do think this can be overcome?
- What is technological consciousness and how does it relate to human thought and action?
- What would be the consequences if artificial entities were treated as autono-mous and responsible moral agents? What rights and responsibilities should they be granted?
- Are the conditions for moral responsibility the same as for legal responsibil-ity or different? Should an artificial entity be held legally liable? If so, under which conditions?

- Is personhood a necessary prerequisite for ascribing moral agency or could moral agency be attributed to artificial entities?

DISCUSSION

This chapter, albeit abstract, plays an important role in "taking back" the richness of technology that has become narrowed through recent conceptualizations of the term. In doing so, it presents a more comprehensive view of technology deeply entrenched in human history, society, and mind. To this end, it attempted to expel common myths about technology and consciousness to provide more suitable base for scholarly inquiry and debate in technology studies and its ethical aspects. It reinforces the idea that at the heart of any serious study of technology and society lies the study of human consciousness that shapes and is shaped by technology and society.

REFERENCES

Ascott, R. (2003). *Telematic embrace: Visionary theories of art, technology and consciousness.* Berkeley, CA: University of California Press.

Bacon, F. (2008). *Francis Bacon: Complete essays.* London: Dover Publications

Bunge, M. (1977). Towards a technoethics . *The Monist, 60*(1), 96–107.

Churchland, P. (1989). *A neurocomputational perspective: The nature of mind and the structure of science.* Cambridge, MA: MIT Press.

Cicero, M. (1972). *The nature of the gods.* Transl. H. McGregor. London: Penguin Books.

Clark, A. (2001). *Mindware: An introduction to the philosophy of cognitive science.* New York: Oxford University Press.

Dennett, D. (1991). *Consciousness explained.* New York: Little Brown.

Dreyfus, H. L. (1992). *What computers still can't do,* (3rd ed.). Cambridge, MA: MIT Press.

Feyerabend, P. (1975). *Against method: Outline of an anarchistic theory of knowledge.* Chicago: University of Chicago Press.

Feyerabend, P. (1978). *Science in a free society.* Chicago: University of Chicago Press.

Floridi, L., & Sanders, J. W. (2004). On the morality of artificial agents. *Minds and Machines, 14*(3), 349–379. doi:10.1023/B:MIND.0000035461.63578.9d

Franklin, S., & Graesser, A. (1996). Is it an agent, or just a program: A taxonomy for autonomous agents. In *Proceedings of the Third International Workshop on Agent Theories, Architectures, and Languages*. Berlin: Springer-Verlag.

Giere, R. (1988). *Explaining science: A cognitive approach*. Chicago: University of Chicago Press.

Gorayska, B., & Mey, J. (2004). *Cognition and technology: Co-existence, convergence, and co-evolution*. Amsterdam: John Benjamins Publishing Company.

Gorman, M. (1992). *Simulating science: Heuristics, mental models and technoscientific thinking*. Bloomington, IN: Indiana University Press.

Heidegger, M. (1977). The question concerning technology. In W. Lovitt, (Ed.), *The question concerning technology and other essays* (pp.13-39). New York: Harper and Row.

Johnson-Laird, P. (1988). *The computer and the mind: An introduction to cognitive science*. Cambridge, MA: Harvard University Press.

Kant, E. (1787). *Critique of pure reason*. Transalted by W. Pluhar & P. Kitcher (1996). Indianapolis: Hacket Publishing.

Kuhn, T. (1962). *The structure of scientific revolutions*. Chicago: University of Chicago Press.

Luppicini, R., & Adell, R. (Eds.). (2008). *Handbook of research on technoethics*. Hershey: Idea Group Publishing.

McClelland, J., & Rumelhart, D., & the PDP Research Group (Eds.). (1986). *Parallel distributed processing: explorations in the microstructure of cognition,* (vol.2). Cambridge: MIT Press.

McDermott, J. J. (1976). *The Culture of Experience: Philosophical Essays in the American Grain*. New York: New York University Press.

Minsky, M. (1985). *The society of mind*. New York: Simon and Shuster.

Mumford, L. (1934). *Technics and civilization*. New York: Harcourt Brace.

Nye, D. (2007). *Technology matters: Questions to live with*. Cambridge, MA: MIT Press.

Ranade, S. (1998). *The technology of consciousness*. Ney York: DIPTI Publications

Simon, H. (1966). *The shape for automation for men and management.* New York: Harper & Row.

Stahl, B. C. (2006). Responsible Computers? A Case for Ascribing Quasi-Responsibility to Computers Independent of Personhood or Agency. *Ethics and Information Technology, 8,* 205–213. doi:10.1007/s10676-006-9112-4

Stanford Encyclopedia of Philosophy. (2008). *Techné.* Stanford, CA: Stanford University Press.

Sullins, J. (2008). Artificial moral agency in technoethics. In R. Luppicini & R. Adell (eds), *Handbook of research on technoethics* (pp. 205-221). Hershey, PA: Idea Group.

Weiser, M. (1999). The spirit of the engineering quest. *Technology in Society, 21,* 355–361. doi:10.1016/S0160-791X(99)00030-5

ADDITIONAL READING

Ascott, R. (1999). *Reframing consciousness.* Exeter: Intellect.

Ascott, R. (2005). *Engineering nature: Art & consciousness in the post-biological era.* Exeter: Intellect.

Friedenberg, J. D., & Silverman, G. (2005). *Cognitive science: An introduction to the study of mind.* Thousand Oaks, CA: Sage.

Giere, R., & Moffatt. (2003). Distributed cognition: Where the cognitive and the social merge. *Social Studies of Science, 33,* 301–310. doi:10.1177/03063127030332017

Goldman, A. (1993). *Philosophical applications of cognitive science.* Boulder: Westview Press.

Luhmann, N. (1996). *Social systems.* Translated by J. Bednarz and D.Baecker. Stanford: Stanford University Press.

Searle, J. (1992). *The rediscovery of the mind.* Cambridge, MA: MIT Press.

Simon, H. (1969). *The sciences of the artificial.* Cambridge: MIT Press.

Thagard, P. (2005). *Mind: Introduction to cognitive science (*second edition). Cambridge, MA: MIT Press.

Chapter 4
Technology Assessment and Technoethics Inquiry

INTRODUCTION

You cannot manipulate the world as if it were a chunk of clay and at the same time disclaim all responsibility for what you do or refuse to do, particularly since your skills are needed to repair whatever damages you may have done or at least to forestall future such damages - Mario Bunge, 197, p. 96.

Philosophical and historical conceptualizations of technology and ethics discussed in the first three chapters of this book helped to situate the reader within the general context of technoethical scholarship as it pertains to a knowledge society. This is important in providing a theoretical grounding for the field of Technoethics. How-

DOI: 10.4018/978-1-60566-952-6.ch004

ever, the field of Technoethics also has an applied research and practice orientation for guiding technoethical inquiry and its application to technology assessment and technology design. As such it offers practical tools for use within the technology oriented professions (e.g., engineering, computer science, medicine, technology studies).

Scientific and technological advances in the 21st century embedded within many facets of society and the lives of individuals are complex, far reaching, and context specific. These advances have the power to greatly improve life on this planet or destroy it forever. Any attempt to provide a theoretical framing for technological inquiry must be able to fit the complexity and scope of current (and future) technological developments. First, technological developments are complex with many stages or phases to take into consideration which map onto their design, development, implementation, and evaluation in society. Social and ethical considerations need to inform a complex array of processes including, reviewing background research and knowledge relevant to technological development, analyzing current research on technological design and development along with relevant codes and policies governing its use, assessing the potential positive and negative contributions along with risks, and evaluating all knowledge gathered in the technological inquiry to guide decision-making. Second, technological development is also broad in scope affecting diverse areas of society with and those within it. New technologies are emerging in communications, transportation, medicine, health, engineering, nutrition, entertainment, and many other areas. Third, technological developments are unique affecting diverse areas of society with different social and ethical concerns arising for those within it. Ethical debates about new industrial technologies may focus on environmental effects with little concern over privacy, while debates around new information and communication technologies may highlight security and privacy issues while other aspects are less relevant. For these reasons, a systems approach to technoethical inquiry is chosen.

Central to technoethical inquiry is the concept of a system (a configuration of parts connected together through by a web of relationships). Technoethics pioneer, Mario Bunge, viewed society as a system of interrelated individuals sharing an environment which could be formalized. Bunge (1979) stated, "A society X is representable as an ordered triple (composition of X, environment of X, components of X), where the structure of X is the collection of relations (in particular connections among components of X (p. 13). Although system complexity (e.g., number of sub-system components, system properties, etc.) varies with the system, the basic ontology of society (and its sub-systems) under s systems view is as follows:

1. A society is neither a mere aggregate of individuals nor a supra-individual entity: it is a system of interconnected individuals.

2. Since society is a system, it has systemic or global properties. Some of these properties are resultant or reducible and others are emergent—they are rooted to the individuals and their interplay but do not characterize them.
3. Society cannot act on its members but the members of a social group can act severally upon a single individual, and the behavior of each individual is determined not only by his genetic make-up but also by the role he plays in society (p. 16)

Common to most (if not all) system theories is the assumption that any complex phenomena can be studied as a system and that any system (mechanical, biological, social) has common patterns and properties that can be explained and modeled using systems methodology (models, strategies, tools) to provide insight into the nature and operation of the complex phenomena under investigation. Many of these and properties (even within simple technological systems) have goal states and values built into them. As stated by Bunge (1977):

We have learned that every control system, be it a furnace with a thermostat or an organism endowed with a nervous system, has an ought built into it in the form of a set of final or goal states which the system tries to attain or to keep. Any such system behaves in such a way that its 'is' approaches its 'ought', the size of its misalignment thus being reduced (p. 102).

To this end, technoethical inquiry, as it is developed in this book, is a social systems theory and methodology concerned with technological relations within society. As will be shown, the systems methodology derived from technoethical inquiry provides powerful tools to help integrate interdisciplinary expertise (and experts) and deal with competing values and ethical issues connected to the actual and potential influences of technology in society.

BACKGROUND

Pragmatism

Early historical groundwork for technoethical inquiry can be traced back as far as John Dewey and Charles Peirce. The social pragmatist elements in Dewey's philosophy are considered a forerunner to technoethical inquiry. Dewey believed the amalgamation of science and technology into human activity would help to make the pursuits of both more intelligent while increasing the level of human control. Dewey did not advance a formal ethical theory of technology to guide public interest in political affairs. He did, however, address the serious need for society to exploit technology to improve communication and public interest in democratic decision making on important issues that influence society and the modern world (Dewey, 1927). In *The Public and its Problems (1927)*, Dewey portrays technology as a distracting force contributing to the lack of public participation in political decision making. This text describes how modern technology, in combination with corporate interests and the ambiguous nature of public communication, can distract individuals from participation in public decision making on important matters of societal concern. Dewey viewed the advent of new technologies in modern society (movies, motor cars, etc.) as a powerful diverting force fragmenting the public into many public spheres with special interests. He believed this diverting force occurred at the expense of public interest in political decision making. This line of reasoning also runs through *Liberalism and Social Action* where Dewey describes ethical inquiry into technology as a form of social action intended to critically examine potential outcomes of technology advancement in human activity (Dewey, 1935).

The methodology of Pragmaticism (What Peirce called pragmatism) is orientated towards practical consequences and real effects as core components of intellectual inquiry into meaning and truth in the world. Pragmatism operates according to a normative principle (pragmatic maxim) formulated by Charles Sanders Peirce to guide intellectual inquiry. As stated by Peirce (1998), "Consider what effects, that might conceivably have practical bearings, we conceive the object of our conception to have. Then, our conception of these effects is the whole of our conception of the object. (CP v. 5, para. 2.). Pragmatism highlights the primacy of practice, operational theory use, the importance of naturalistic inquiry, and the inclusion of values (ethics) and fact within inquiry.

Technoethoical inquiry adheres to core pragmatist assumptions. It does not focus on ethical theories independent of experience. Instead it views theory and practice as emerging from experience (and existing knowledge) within the environment (primacy of practice). This entails an operational approach to theory development where a theory proves itself successful and closer to the truth if it predicts or controls something in the world better than alternative theories available. This operational approach to theory development and and truth seeking is employed by scientists and technologists. This fits the problem context of many technology concerns arising from real life trial-and-error based professional activities and practices in various areas of society. Technoethical inquiry is aligned with naturalistic inquiry which treats human knowledge as something that can be studied empirically using scientific (and other) methods. This situates technoethical inquiry within the realm of human experience and social practice while placing technoethical inquiry firmly in the hands of those working with and affected by technology. Moreover, echnoethical inquiry follows pragmatism in not creating a dichotomy between fact and value. In other words, knowledge describes what individuals believe and this includes empirical facts and values concerning what is good and not good.

A number of interesting pragmatist tools are available to study technological systems such as, case-based reasoning and casuistry. Case-based reasoning is a pragmatist tool for problem-solving based on the solutions of similar past problems. This tool is used in many technology oriented problem contexts. For instance, a computer technician may repair a computer by recalling another computer that demonstrated similar problems. A lawyer fighting for a computer designer's patent rights may advocate for a particular trial outcome based on legal precedents set in previous cases. Case-based reasoning is pervasive in technology oriented professions and in everyday human problem solving (Aamodt, Agnar & Plaza, 1994).

Casuistry is a specialized method of case reasoning for dealing with ethical aspects of cases by applying moral rules to particular instances where there are conflicting values or when unique circumstances alter cases. In terms of application, casuistry begins by examining an actual case and drawing parallels between this case and a pure case (paradigm or type case). Based on this comparison, a moral response can be applied to the particular case. If the actual case is identical to the type case, moral judgments can be offered using moral principles from the type case. However, if the actual case differs from the type case, then the differences can be critically assessed. Jonsen and Toulmin's (1988), *The Abuse of Casuistry: A History of Moral Reasoning* (1988) effectively applied casuistry in practical argumentation by using type cases as referential markers in moral arguments to make rational claims to address differences between actual and type cases. Under Toulin's practical argumentation model, good practical arguments can be used to justify claim, which can withstand criticism (Toulmin, 1958). Toulmin (1958) provided a theoretical framework for analyzing practical arguments reworked into the following dimensions and guiding questions:

- **Claim:** Has the merit of the conclusion been established?
- **Evidence:** Do the facts appealed to support the claim made?
- **Warrant:** Do statements made justify a link between the evidence and claim?
- **Backing:** Are credentials provided to back statements of warrant?
- **Rebuttal:** Do statements acknowledge any restrictions to which a claim may be legitimately applied?
- **Qualifier:** To what extent do statements express certainty or conviction about the claim?

What is interesting in the application of this pragmatic tool (casuistry) is that it helps to highlight how technoethical inquiry applies to technology problem contexts. First, knowledge about the phenomenon under investigation is derived from naturalistic observation within practice (i.e., providing case details, background information, influencing factors). Next, operational theory is used (constructing type case, comparing and ranking moral values) to generate theoretical knowledge from facts and values interpreted from the case. Finally, additional tools (Toulmin's practical argumentation model) was introduced to provide reasons for ethical decision-making that can be justified and accepted by social system members. This is important in pragmatist methodology. As stated by Peirce (1998), "The opinion which is fated to be ultimately agreed to by all who investigate, is what we mean by the truth, and the object represented in this opinion is the real. That is the way I would explain reality" (pp. 139).

The Ethical Turn in Technological Inquiry and Assessment

The need for an ethical orientation to technological inquiry began materializing in the 1960s. Seminal contributions to technoethical inquiry derive from works such as Carson's (1962) *Silent Spring*, Schumacher's (1973) *Small Is Beautiful*. Carson's (1962) *Silent Spring* was a pioneering environmentalist study focusing on the environmental impacts of chemical (such as DDT) to raise awareness that technology can destroy the earth's environment and threaten the quality of human life. Carson asserts, "The most alarming of all of man's assaults on the environment, is the contamination of air, earth, rivers, and sea with dangerous and even lethal materials" (6). This critical work helped raise awareness and public interest about the destructive power that technology can exert on human life and the environment. Schumacher's (1973) *Small Is Beautiful* was a driving force behind the 'appropriate technology movement.' This text argued that traditional (low-tech) technologies can often result in greater overall productivity compared to expensive high-tech methods, while minimizing other costs including social dislocation. The author reviews economic and technological cases to show how modern society has unwisely invested resources in creating larger and more powerful technologies to benefit the economy without regard to other costs, including environmental costs:

Ever bigger machines, entailing ever bigger concentrations of economic power and exerting ever greater violence against the environment, do not represent progress: they are a denial of wisdom. Wisdom demands a new orientation of science and technology towards the organic, the gentle, the non-violent, the elegant and beautiful (107).

Other works under the heading risk assessment and risk management continue this diagnostic approach to evaluating technology in terms of its outcomes. Perrow's (1984) *Normal Accidents: Living with High-Risk Technologies* and Beck's (1992) *Risk Society: Towards a New Modernity* are excellent studies in the art and practice of technology assessment. Perrow (1984) examines the social and ethical aspects of technological risk using real life situations to illustrate the shortcomings of conventional engineering approach to ensuring safety due to the complexity of modern technical systems. The text warns that "Risk will never be eliminated from high-risk systems and we will never eliminate more than a few systems at best" (4). Beck's (1992) *Risk Society: Towards a New Modernity* extends the social and ethical challenges of technological risk assessment which tend to increasingly favour the wealthy stakeholders in society. The text states, "As the risk society develops, so does the antagonism between those afflicted by risks and those who profit from them. The social and economic importance of knowledge grows similarly, and with it the power over the media to structure knowledge and disseminate it" (46). This expanding body of scholarship in technological inquiry and later risk assessment approaches highlights the broader social and ethical influences that underlie technological risk. This work also provides a useful knowledge building tool for technoethical inquiry on the potentially negative outcomes of technological innovation.

CONTEXT OF TECHNOETHICAL DEPLOYMENT

Technoethical Inquiry as a Social System Theory and Methodology

How do we guide technology in an ethical way to leverage society? How can technological innovation and human innovation be reconciled within an ethical framework? Technoethical inquiry provides a unique conceptual framework for studying the ethical and social context of science and technology in society. How is this accomplished? First, technoethical inquiry is based on a broad conceptualization of technology as a system closely connected to human activity in society. This highlights the centrality of knowledge creation situated within scientific and technological structures embedded within individual life, society and the environment. Technology viewed in a broad sense of the term is important for multiple reasons. It avoids existing work that defines technology in overly narrow terms which fail capture all the varieties of current (and new) technology present in the world. It assumes that any conceptualization of technology must be specific enough to inform individual technologies and robust enough to address general aspects of technology as a system entrenched in an evolving technological society.

Second, system structures and relational operations are dynamic and constantly in flux due to changing system relations between technology, nature, and society.

This system can be understood by appreciating the various areas of system operations: biological, environmental, individual, and social. On an environment level, technoethical inquiry explores the ethical aspects of science and technology in the world around us. For instance, to what extent do environmental studies, stakeholder interests, and expert knowledge shape views on the environment in the midst of technological innovation? To what extent do ethical considerations about technology influence technological innovation that affects the environment? On a biological level, technoethical inquiry explores the ethical aspects of science and technology used to influence biological and physical systems. For instance, to what extent do studies of biological and physical systems, stakeholder interests, and expert knowledge shape views of our bodies (including genetics, nutrition, reproductions, etc) in the midst of technological innovation? To what extent do ethical considerations about technology influence biological and physical processes associated with technological innovation? On a societal level, the technical inquiry focuses on the ethical aspects of science and technology used to influence social systems. For instance, to what extent do societal studies, stakeholder interests, and expert knowledge shape the view social identity (including identity building, social interactions, community involvement, etc) in the midst of technological innovation? To what extent do ethical considerations about technology influence societal processes associated with technological innovation? On an individual level, the technical inquiry focuses on the ethical aspects of science and technology that affect individual consciousness and mental operations. What is the role played by technological consciousness within the human mind? To what extent do cognitive studies shape views of individual identity in relation to technological creations? To what extent do ethical considerations about technology relate to technological consciousness and technological processes?

Third, humans are part of technology as a system rather than being independent of it. This sense of technology places it in close relation to human activity including technology assessment, technology design, reflective practices, knowledge production (facts and values), and group decision-making. The human relationship with technology is at the forefront of inquiry because it is only by examining our relationships with technology in it various forms and contexts that humans acquire the necessary knowledge to make informed decisions about technology. This rejects the positivist notion of humans as neutral observers independent of technology. Instead, it places humans and technology within a reflexive system that questions itself and adapts to changing relations.

Technoethical inquiry is a social system theory of technology and methodology for studying technological systems. The system parts and resulting web of relations (technical and social) are involved in goal directed behaviors to generate technological knowledge within society (phenomena). Following Varela (1981), social systems are autopoietically closed systems (self-producing), maintaining system identity and operations while relying on resources from the environment. Technological systems, as social systems, are self-producing systems of technological knowledge creation (facts and values) within the environment. Knowledge creation in this system operates by processing meaning derived from previous knowledge and system operations (experiences). The identity of this system is constantly produced and reproduced through multiple forms of knowledge creation (i.e., ideas, actions, normative rules, artifacts) about itself and what is considered meaningful within the system. If some system function fails to maintain itself (as often happens when old technologies and forms of technological knowledge become obsolete in society) it ceases to exist.

Technology, as a system, is a complex and adaptive system. Multiple forms and multiple types of knowledge are created and recreated within this system. This is because humans (and technology) are positioned within this social system, which derives knowledge and meaning from human consciousnesses, human activities, social practices, and social norms and values. The multi-faceted nature of knowledge creation within this system provides a variety of possible operations by which this system can adapt to the environment and fulfill functions which contribute to our evolving knowledge society.

The Aims of Technoethical Inquiry

The overall aim in technoethical inquiry is to advance technoethical knowledge and guide ethical actions related to the technological phenomena studied. The following are posited as guiding principles of technoethical inquiry:

1. Technoethical inquiry treats technology as a self-producing social system that re/produces itself on the basis of knowledge creation (facts and values). This self-producing system of technology is the core of the evolving knowledge society.
2. The derivation of meaning about system operations is situated in various areas of human activity and society affected by technology. As such, it involves multi-perspective and multi-aspect technoethical inquiry that uses interdisciplinary expertises to addresses various aspects of system operations (environmental, biological, individual, and social relational operations) where technology plays a role.
3. The outcome of a successful technoethical inquiry is defined as the point at which a shared understanding/knowledge is demonstrated concerning relevant ethical aspects of a set of technological relations with no new knowledge emerging. This does not require consensus, but rather the identification of all relevant knowledge (facts and values) and priorities (value ranking) applicable to the technological relations to which a technolethical inquiry is applied.

These three defining features are basic guidelines to be applied when framing technoethical inquiry. To conclude, Technoethics is studied as a complex multi-level system where shared understanding emerges from knowledge creation embedded within multi-level system operations and the ethical tensions that arise from the technological phenomena under investigation. How do ethical tensions within this system get resolved? Technology, as it is treated in technoethical inquiry, is a core system relation deeply connected to the environment, life, and human interest within an evolving society. As such it is a reflective system that humans can study and learn from. Acquiring knowledge about technology is among the technologies that constitute this system and one of the powerful tools that humans possess. This is because knowledge creation is a tool that allows individual to acquire knowledge and understanding about technology and its ethical aspects. This, in turn can used to shed new light on positive and negative elements of technology to help guide ethical decision making. A successful technoethical inquiry provides opportunities for humans to better understand ethical responsibilities created by technological in-novations. This, in turn, can be used to steer the system in ways that reduce ethical tensions within it. Key applications of technoethical inquiry to be discussed include technoethical assessment (TEA) and technoethical design (TED) .

In terms of research tools, technoethical inquiry is not limited to any one set of data collection and analysis strategies. Rather, it uses a variety of research approaches and techniques drawn from technical studies and the social sciences. This includes but is not limited to needs analysis, systems research, archival research, content analysis, semiotic analysis, case study, historical research, interviews and focus groups, life history, narrative research, observation studies, ethnographic research, participant observation, survey research, evaluation studies (formative and summative), us-ability testing, and longitudinal studies. The breadth of approaches and techniques available for guiding technoethical inquiry is illustrative of the interdisciplinary nature of Technoethics and the group of scholars working within it.

TECHNOETHICAL ASSESSMENT (TEA)

What can technology assessment contribute to our understanding of technology and ethics in society? Technology Assessment (TA) describes the evaluation of new technologies within society. Techoethical Assment (TEA) assumes that the assessment of technology cannot be limited to scientific evaluation but must be sensitive to broader social and ethical elements within society. This ethical turn in technology assessment is reflected in current work in the field. Mohr (1999) lists the following commonly held beliefs among TA experts:

- If we want to gain the future we depend on technological progress
- We are aware of the ambivalence of any technological progress
- There are no simple answers with regard to progress in a pluralistic world where preferences and aims are controversially disputed
- We are as responsible in an ethical sense for what we are doing as we are responsible for what we are not doing: to act or not to act— principle of ethical equivalence
- Technology assessment is obliged to analyze and evaluate the desirable and the non-desirable consequences, the chances and the risks, of technologies, new technologies as well as established technologies. At present the major threats to the future stem from firmly established technologies such as burning fossil materials, not from novel technologies.
- The motto of TA is that a new technology must be better than the preceding technology. Otherwise we do not need it. "Better" does not only refer to the scientific evaluation of a technology but also to the social (socioeconomic) and environmental dimensions (p.2).

The ethical orientation in technology assessment is reflected in contemporary technoethical assessment scholarship. Building on Bunge (1977), technoethical assessment assumes a pragmatic orientation to moral norms grounded in human activity and existing knowledge (facts and values). Bunge (1977) explains, "Instead of accepting rules of thumb in the realm of morals we can and should try to form them in the image of technological rules, i.e., on the strength of factual knowledge and objective valuation" (p. 106). Technological rules (moral and technical) are derived through the ranking of value judgements (interests) and knowledge statements pertaining to the focus of a technoethical inquiry. Minimally this requires an analysis of the means, intended ends, and possible side effects subjected to moral and social controls (i.e., stakeholder interests). The following are basic steps for creating technological rules in a minimal detail context:

- **Step 1:** Evaluate the intended ends and possible side effects to discern overall value (interest)
- **Step 2:** Compare the means and intended ends in terms of technical and non-technical (moral, social) aspects
- **Step 3:** Reject any action where the output (overall value) does not balance the input in terms of efficiency and fairness

However, most (if not all) problem contexts for technoethical inquiry are more complex and require additional steps and considerations. Such is the case when dealing with any large-scale technology initiatives which impact society and the environment (e.g., establishing new mining operations, factory building, creating new pharmaceutical drugs, creating robots to aid in the workforce). In such cases, multiple stakeholder interests must be integrated into the technological inquiry (Steps 1-3) in order to ensure that all perspectives are represented from groups affected by the outcome of technological initiative if brought to fruition. This warrants the addition of a fourth step:

Step 4: Technoethical inquiry is multi-perspective. Ensure that perspectives from all stakeholders groups and those affected are included in Steps 1-3

This does not entail equal participation from all groups affected because this may vary. Neither does it limit how such multi-perspectives are integrated since there are a variety of tools at the disposal of the technoethicist (e.g., conversation modelling, interview and focus group research, survey methods, group decision-making frameworks, rational argumentation models, conflict resolution frameworks, etc). It does require, however, that all groups affected have the opportunity to participate. This is particularly important when entire communities can be negatively impacted. Following Bunge (1977):

The community affected by the project has the right to keep it under the control of other specialists—applied social scientists, public health officials, city planners, conservationists, etc—to the point of vetoing the whole thing if its negative effects are likely to outweigh its social benefits. It is a question not of slowing down technological progress but of preventing progress in one respect (e.g., engineering design) from blocking advancement in other respects (p.101).

The other core concern revolves not around who is affected but rather, what is the scope or level(s) at which some technological project has an impact (I.e. biological, physical, psychological, social, environmental). This gives rise to a fifth step:

Step 5: Because technological projects can have multi-level influences (i.e., biological, physical, psychological, social, environmental), technoethical inquiry must consider technological relations at a variety of levels

As a complex social system theory and methodology, Technoethics requires a mode of inquiry that matches or models the complexity of the system. Steps 4 and Step 5 are closely connected and help to provide additional strategies to help ensure the validity of technological rules used to guide action. The complexity introduced by acknowledging multiple levels of technological influence (Step 5) create a complex set of technological relations that cannot be solved by a single group of experts (Step 4), even if those experts are trained engineers or other technology professionals. As noted by Bunge (1977):

Because of the close relationship among the physical, biological and social aspects of any large-scale technological project, advanced large-scale technology cannot be one-sided, in the service of narrow interests, short-sighted, and beyond moral control: it is many-sided, socially oriented, far-sighted, and moral bridled (p. 101).

TECHNOETHICAL DESIGN

As discussed in previous sections, the overall aim in technoethical inquiry to advance technoethical knowledge and guide ethical actions related to the technological phenomena studied. An important part of this lies in the design of technological systems. As a part of technoethical inquiry, technoethical design (TED) guides technological systems design and development with a focus on ethical and social aspects of system operations. Thus, the goal of technoethical design is to create optimal systems and optimize existing technological system operations to meet the needs of all affected. This requires an understanding of the technological system and how it works. Bunge (1979) noted that optimizing any social system requires an understanding of essential social system components and operations:

The very first thing to do is to identify the components, the environment and the structure of the system. A second step would be to attempt to disclose the state variables of the system—at the very least its inputs and outputs. A third step may consist in hypothesizing definite relations among the state variables, and a fourth in either simulating these assumptions on a computer or putting them to an empirical test (p. 29).

Technoethical system design (or redesign) requires design knowledge of system parts and operations. Technological systems are social systems with humans (with changing interests and needs) positioned within this system. As such, technology is a complex and adaptive system where multiple forms and multiple types of knowledge are created and recreated. This creates a number of challenges connected to technoethical design efforts that need to be addresses. First, To understand technoethical design is to understand the complexity of technology as a social system with moral aspects. As noted by Verbeek (2006):

The ethical ambition to design technologies with the explicit aim to influence human actions raises moral questions itself. It is not self-evident, after all, that all attempts to steer human behavior are morally justified, and steering human beings with the help of technology raises associations with the totalitarian technocracy of Orwell's Big Brother (p. 363).

Unlike pure science which focuses on the pursuit of knowledge as a good in itself, technological systems are intertwined within human values and activities which affect life, society, and the environment. As stated by Bunge (1977), "Technology gives power over things and men—and not all power is good to everyone" (p. 100). For this reason, technologists have the dual task of attending to technical and moral aspects of technology. Because technoethical design is concerned with the interplay of human actions (and values) and technology, design knowledge derives from multiple perspectives including designers, ethicists, stakeholders, and the public affected (or potentially) affected. This requires a high level of organization and the willingness to engage in participatory design efforts needed to inform system design and development.

What further complicates technoethical design efforts lies in the elusive nature of technology relations in society. This can be expressed in terms of technology mediation. Technology mediation deals with the role technology plays in shaping or steering human-world relations. As stated by Verbeek (2005):

On the one hand, the concept of mediation helps to show that technologies actively shape the character of human–world relations. Human contact with reality is always mediated, and technologies offer one possible form of mediation. On the other hand, it means that any particular mediation can only arise within specific contexts of use and interpretation (p. 11).

Within the context of technoethical design, technology mediation is a relational process concerned with how technology mediates and is mediated by human experiences and interpretations of technological relations.

Because of the ethical aspects of technology-human relations, the mediating role of technology must be integrated in the design process. The challenge is to make the hidden forms of technology mediation (see chapter 2) explicit to all to create connections between the design context and the context of use. This requires attention, not only to the contextual specificity of technology mediation but also its possible consequences. As reflected in Verbeek (2006):

To build in specific forms of mediation in technologies, designers need to anticipate the future mediating role of the technologies they are designing. And this is a complex task since there is no direct relationship between the activities of designers and the mediating role of the technologies they are designing (p. 372).

Based on the abovementioned deign considerations, the following basic steps for guiding technoethical design (TEA) are offered:

- **Step 1:** Make sure the technological system is explicitly understood by all members within the design context. This includes system components and relations such as forms of technology mediation (and their impact).
- **Step 2:** Perform technoethical assessment to discern all possible technical knowledge (facts and values) relevant to the design context.
- **Step 3:** Focus design efforts on optimizing technological system operations to satisfy the needs and interests of all affected.
- **Step 4:** Consult with representative from all groups affected by some technology in an effort to establish mutual understanding and agreement concerning important design issues.

In terms of design step orientation, Steps 1 and 2 focus on the acquisition of knowledge concerning the unique components and relations that constitute some technological system for design (or redesign). It emphasises the inclusion of both values and facts in system design and redesign processes (step 1). In other words, values are part of the normal valuation function of systems. Bunge (1977) states, "Once values are recognized as the outcome of a valuation activity of an organism [system], they cease to be disjointed from facts: they become aspects of certain facts" (p. 102). This valuation aspect of system functioning is particularly important within technoethics, as a social system embedded within a knowledge society reliant on knowledge creation (including facts and values). It also involves attention to underlying context and future possibilities connected to the technological system design. Verbeek (2005) remarked, "To understand the role of technology in human existence, one must think not only backward to its conditions of possibility, but also forward to what technological artifacts themselves make possible, and what this means for human existence"(p. 30). Step 3 and 4 highlight the shift in contemporary technology systems design from an isolated technologist led design process to a multi-perspective and user focused design process that takes a broader focus on user interests and needs that goes beyond technical aspects. As sated by Bunge (1977), "The technologist must be held not only morally responsible for whatever he designs or executes: not only should his artifacts be optimally efficient but, far from being harmful, they should be beneficial, and not only in the short run but also in the long term (p. 100). This type of activity requires active engagement and consultation with the users and potential users (step4). As indicated by Verbeek (2006):

Technologies have to be interpreted and appropriated by their users to be more than just objects lying around. Only when human beings use them, artifacts become artifacts for doing something. And this "for doing something" is determined not entirely by the properties of the artifact itself but also by the ways users deal with them (p. 372).

It is worth nothing that the above steps are not intended as rigid procedures to be generalized to all technology design contexts. Rather, they are guiding principles or rules of thumb to be modified according to designer needs, technological system scope, context specificity, and other design constraints (e.g., time, money, available expertise, etc.). It also must be noted that not all technological systems contain equally potent ethical aspects which require all design steps to be followed. Grunwald (2001) offers the following criteria that designers can use in determining whether a technoethical inquiry is needed or whether it is unnecessary because an already well established normative framework is in place with the following characteristics:

- **Pragmatically complete:** The normative framework has to comprehend adequately the decision to be made and should leave out no essential aspects from consideration;
- **Locally consistent:** There has to be a sufficient degree of freedom from contradiction among the various elements of the normative framework;
- **Unambiguous:** Beyond the normative framework, there has to be a sufficient common understanding among the actors in the context of the decision under discussion;
- **Accepted:** The normative framework has to be accepted as the basis for the decision by those concerned;
- **Observed:** The normative framework has to be in fact observed, and lip-service, for instance, in environmental concerns, is not enough (Grunwald 2001, p. 419).

Thus, technoethical design principles may be selectively applied and molded in different degrees depending on the design context. In some cases, it many be streamlined. Such is the case when dealing with minor design modifications to well entrenched and tested technological systems or systems which have relatively minimal impact on humans and society.

DISCUSSION QUESTIONS

- Some countries have established agencies to help guide ethical decision making concerning technological innovation. Do you think such agencies are needed?

- What barriers do you think a national or international agency to regulate ethical innovation would confront within the current political climate?
- What does technology assessment with an ethical focus contribute to effective technology innovation and protect against negative consequences?
- How does social systems theory approach the study of society? What are the advantages of technoethical inquiry as a systems theory of society?

DISCUSSION

Technoethics, as it is advanced in this chapter, is grounded in a systems theory and methodology for guiding technoethical inquiry. Under this framework, technoethical inquiry seeks to understand technological phenomena as sets of system relations situated within the environment. The chapter explored how systems methodology provides powerful tools to help integrate interdisciplinary expertise (and experts) and deal with competing values and ethical issues connected to the actual and potential influences of technology in society. The chapter applied principles of technoethical inquiry to applications in technology assessment and technology design in an effort to highlight the practical's implications of adapting Technoethics.

REFERENCES

Aamodt, Agnar, A., & Plaza, E. (1994). Case-based reasoning: Foundational issues, methodological variations, and system approaches. *Artificial Intelligence Communications*, 7(1), 39–52.

Beck, U. (1992). *Risk society: Towards a new modernity.* London: Sage.

Bell, D. (1973). *The coming of post-industrial society: A venture in social forecasting.* New York: Basic Books.

Bunge, M. (1977). Towards a technoethics. *The Monist*, 60(1), 96–107.

Bunge, M. (1979). A systems concept of society: Beyond individualism and holism. *Theory and Decision*, 10(1), 13–30. doi:10.1007/BF00126329

Carson, R. (1962). *Silent spring.* Boston: Houghton Mifflin.

Dewey, J. (1927). *The public and its problems.* New York: Holt.

Dewey, J. (1935). *Liberalism and social action.* Carbondale, IL: Southern Illinois University Carbondale.

Feenberg, A. (1991). *Critical theory of technology.* New York: Oxford University Press.

Grübler, A. (2003). *Technology and global change.* Cambridge, UK: Cambridge University Press.

Grunwald, A. (2001). The application of ethics to engineering and the engineer's moral responsibility: Perspectives for a research agenda. *Science and Engineering Ethics, 7*(3), 415–428. doi:10.1007/s11948-001-0063-1

Jonsen, A., & Toulmin, S. (1988). *The abuse of casuistry.* Berkeley, CA: University of California Press.

Mohr, H. (1999). Technology Assessment in Theory and Practice. *Techné: Journal of the Society for Philosophy and Technology, 4*(4), 1–4.

Peirce, C. S. (1998). *The essential Peirce: Selected philosophical writings (translated 1878).* Bloomington, IN: Indiana University Press.

Schumacher, E. (1973). *Small is beautiful.* New York: Harper & Row.

Toulmin, S. (1958). *The uses of argument.* Boston: Cambridge University Press.

Varela, F. (1981). Autonomy and autopoiesis. In G. Roth & H. Schwegler (Eds.), *Self-organizing systems* (pp. 14-23). Frankfurt: Campus Verlag.

Verbeek, P. (2005). *What things do: Philosophical reflections on technology, agency, and design.* Trans. R. P. Crease. University Park, PA: Pennsylvania State University Press.

Verbeek, P. (2006). Materializing morality: Design ethics and technological mediation. Materializing Morality. *Science, Technology & Human Values, 31*(3), 361–380. doi:10.1177/0162243905285847

ADDITIONAL READING

Bijker, W. (1995). *Of bicycles, bakelites, and bulbs: Toward a theory of sociotechnical change.* Cambridge, MA: MIT Press.

Collins, H., & Pinch, T. (1998). *The golem at large: What you should know about technology.* New York: Cambridge University Press.

Cutcliffe, S. (1989). The emergence of STS as an academic field. *Research in Philosophy and Technology, 9,* 287–301.

Harris, C. E., Pitchard, M. S., & Rabins, M. J. (2000). *Engineering ethics. Concepts and cases*. Belmont, CA: Wadsworth.

Hess, D. (1997). *Science studies: An advanced introduction*. New York: New York University Press.

Meadows, D., et al. (1972). *The limits to growth: A report to the club of Rome's project on the predicament of mankind*. New York: Universe Books.

Roco, M. C., & Bainbridge, S. (2002). *Societal implications of nanoscience and nanotechnology*. Dordrecht, the Netherlands: Kluwer Academic Publishers.

Schummer, J. (2004). Interdisciplinary issues in nanoscale research. In D. Baird, A. Nordmann, & J. Schummer (Eds.), *Discovering the nanoscale* (pp. 9-20). Amsterdam: IOS Press.

Seemann, K. (2003). Basic principles in holistic technology education. *Journal of Technology Education, 14*(2), 12–24.

Shrader-Frechette, K. (1997). Technology and ethical issues. In K. Shrader-Frechette & L. Westra (Eds.), *Technology and values* (pp. 25-32). Lanham, MD: Rowman & Littlefield Publ.

Simon, J. (Ed.). (1995). *The State of Humanity.* Cambridge, MA: Blackwell, 1995.

Strijbos, S., & Basden, A. (Eds.). (2006). *In Search for an Integrative Vision for Technology*. Dordrecht: Springer.

Wajcman, J. (1991) *Feminism Confronts Technology.* University Park: Penn State University Press.

Winner, L. (1984). *The whale and the reactor: The search for limits in an age of high technology.* Chicago: University of Chicago Press.

Chapter 5

The Nature and Areas of Technoethical Inquiry

INTRODUCTION

There is nothing inherently wrong with science, engineering, or managing. But there can be much evil in the goals which either of them is made to serve, as well as in some of the side effects accompanying the best of goals. If the goals are evil—as is the case with genocide, the oppression of minorities or nations; the cheating of consumers, the deception of the public, or the corruption of culture—then of course whoever serves them engages in evildoing, even if not legally sanctioned as wrongdoing.--Bunge, 1977, p. 98.

DOI: 10.4018/978-1-60566-952-6.ch005

As illustrated in the preceding chapters, social and ethical concerns about technology are multifaceted and cannot be resolved through methods derived from any one discipline. Instead, a multi-tiered approach that draws on an interdisciplinary knowledge base is recommended to guide a proper technoethical inquiry advanced through knowledge and insights derived from multiple disciplines and literatures. This approach is desirable for achieving a more comprehensive picture of technology at the core of human life and society. Knowledge derived from the cross-fertilization of relevant areas of inquiry represents a potentially powerful set of knowledge building tools that can be used for maximizing the positive and minimizing the negative ethical aspects of technology in society. To this end, a systems approach to technoethical inquiry (chapter 4) was highlighted as an ideal methodology for studying the multi-faceted nature of ethical aspects of technology. This, however, does not negate the use of other methods and tools available to guide technoethical inquiry. Neither does it capture the nature and scope of technoethical inquiry within the real world of technology and humans.

Regardless of the selected methodology for guiding technoethical inquiry, there is a need to have a variety of supporting tools given the complexity and depth of technological advancement in society. In other words, there is no simple recipe or model for success, but rather a set of tools that can be brought into the field and used when the context demands it. Thus, the aim of this chapter is situate technoethical inquiry within the multi-faceted context of technology and human society. This is accomplished in two ways. In the first section, the multi-faceted nature of technological inquiry is described in terms of the multiple levels (meta-technology, explicit technology, tacit technology) of technological system relations in society. This helps to address the variety of ways that technology manifests itself within society. The multi-faceted nature of technology is also described in terms of key perspectives relevant to the study of technology which warrants interdisciplinary knowledge and expertise. Key perspectives (i.e., philosophical, historical, political, economic, cultural, legalistic) are discussed in efforts to achieve an interdisciplinary focus when studying technological systems. The second section presents the main areas and issues in contemporary technoethical inquiry. This helps familiarize readers with the main topics pursued within the field of Technoethics.

THE MULTI-FACETED NATURE OF TECHNOLOGICAL SYSTEMS

Multiple Levels of Technological System Relations

The key to understanding technology is not about understanding what it is only in terms of technical considerations, but rather, how it connects to life and society by examining the variety of technological system relations possible. This section provides details on the scope of technology as a system which must be appreciated to understand how it connects with technoethical inquiry and its methods of deployment. Following the systems theory and methodology of technethical inquiry (chapter 4), a pragmatic approach to human-technology relations is concerned with how human beings act within society. In other words, when a technology is used, it affects how humans engage with the world and each other (and vice versa).

Within technoethical inquiry, technological systems are core mediators of human-world relationships with a transforming role within a knowledge society. This is because a knowledge society is heavily dependent on technology for key structure and functions. It is also because a technological system and its operation has a transforming power and active role within evolving relationships between humans and their world. In other words, technological mediation has the capacity to augment certain aspects of reality while attenuating other aspects. For instance, observing a hockey game from back row does not allow spectators to see the puck and the letters on players' jerseys directly that is possible from middle to front row seating where some visual details and sounds are discernable (e.g., the smell of sweat from the team benches, the sound of coaches yelling out plays, etc). The use of binoculars or a large overhead screen to broadcast plays allow spectators to pick out such details not discernable with the naked eye even from the middle row seating area. At the same time, some aspects of the hockey game lost (e.g., the smell of sweat from the team benches, the sound of coaches yelling out plays, etc) are lost from direct experience.

A plethora of human-technology relations exist which can be divided into a general taxonomy of technological relations. Figure 1 provides an illustration of the levels of technology found within a knowledge society.

As illustrated in Figure 1, the level of technology refers to the level of abstraction from which the observer views parts of the technological system under consideration. Each level of technology describes how technology mediates human life and society.

Technology that is fairly easy to express and communicate is called *explicit technology* and can be characterized as technology that extends human powers in various ways through its operation. Mumford's (1934) *Technics and Civilization* described the history of explicit technology from the Middle Ages to the 1930s. This work focused on the development of modern technology and how it both shapes and is shaped within evolving civilizations in relation to moral, economic, and political factors. Mumford (1934) stated, "Whereas in industry the machine may properly replace the human being when he has been reduced to an automaton, in the arts the machine can only extend and deepen man's original functions and intuitions" (p. 343). Within any technological system, the knowledge of explicit technologi-

Figure 1. Technology levels within a knowledge society

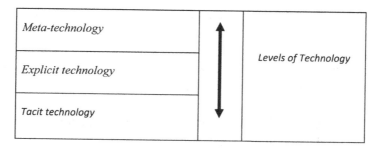

cal operations (including knowledge derived from meta- and tacit technological operations that becomes explicit) occurs at the expressed level of technological abstraction. For instance, a traffic light mediates human action by signaling drivers to stop at intersections when red and to continue driving when green. A traffic light does not provide a direct sensation of oncoming traffic but does mediate traffic by providing a light signal that requires interpretation from the drivers (and pedestrians). A sundial establishes a human-world relationship in terms of time. The sundial reading does not create a direct sensation of time but mediates time by providing a value that requires interpretation from someone to say something about reality (explicit technology).

Tacit technology is at the lower level of abstraction and describes concrete technological system operations which are invisible or implicit to the observer. The invisible and implicit effects of technology (media) were clearly articulated by McLuhan and Parker (1969), "Media effects are new environments as imperceptible as water is to a fish, subliminal for the most part" (22). Tacit technology is widespread in the world of advanced ICT's and "ubiquitous computing" due to flexible technology mediated communications using electronic technologies (i.e., Instant messaging and other chat tools, electronic mail and voicemail, Internet, audio and video conferencing tools, personal data assistants, mobile devices). Electronic technologies provide users with opportunities for flexible interaction, multitasking, and memory storage of digital data (texts, photos, film, music) using social networking tools and digital repositories (i.e., twitter, facebook, my space, flicker). Human use of ICT's surpasses cognitive abilities to consciously monitor all communications in what can be described as an information bomb. Other technologies (biosensors, smart homes) allow users to control aspects of their bodies and environment without having to consciously monitor changes. In other words, this level of human-technology relation is often implicit and unconscious for the user even though it is part of observable human experience. This connects with seminal work on "tacit knowing" (Polanyi, 1966). Polanyi (1966) believed that tacit knowledge was knowledge that individuals carry in their minds that is difficult to access and share. Meta-technology is placed at the level of abstraction above explicit and tacit technology in Figure 1. Meta-technologies operate outside the limits of human experience and sensation (e.g., nuclear energy, nanotechnology, etc.).

Meta-technologies can be characterized as technologies which go beyond human powers by transcending existing technology (explicit, tacit). As stated by Valenella (1999), "Meta-technology thus designates that which transcends and/or supersedes technology by being beyond it, thus modifying the status, situation, and meaning of what is ordered and made intelligible by the logos of technology" (p. 412). Meta-technologies include physical technologies (i.e., instrumentation, created artifacts) and intellectual technologies (i.e., criteria for intelligibility, organizational norms, behavioral rules). Valenella used the concept of meta-technology to describe a complex class of technologies that reorder nature by operating outside the natural order defined by space and time. He characterized this technological-human relation as trans-optical (not intelligible through observation) and trans-human (not intelligible through direct experience). For the purposes of this book, meta-technology refers to technologies whose mediating role can only be understood indirectly through interpretation rather than through natural observation. This is because they operate outside the limits of human experience and sensation within which tacit and explicit technologies operate. Examples of meta-technologies in use today are widespread (e.g., military weapons which use thermal devices to locate (see) targets, the use of radioactive glucose injections in positron emission scanners to allow doctors to see inside human beings, nanotechnologies which operate at a molecular level not observable by the human eye, etc.).

Understanding the levels of technology helps focus technoethical inquiry which seeks to advance knowledge of ethical aspects of technology within a knowledge society. To expand on a term used by McLuhan--Technology is the message. The levels of technology represent different forms of technological system operations which influence the organization of social reality through their mediating role. It acknowledges that technological system operations are more extensive than observers are aware and technoethical inquiry must direct attention to uncovering tacit- and meta-technological operations by transforming them into explicit technological operations to leverage knowledge needed to inform technoethical inquiry. The above model is a visual representation illustrating the multi-faceted nature and complexity of technological systems and is intended as an aid to leverage knowledge acquisition about the nature of human-technological system relations by showing how various technologies contribute differently to life and society as a function of their mediating roles in the development of human-technological relations.

Multiple Perspectives in the Study of Technological Systems

A systems approach to technoethical inquiry provides a theoretical framework for integrating interdisciplinary expertise to address the plethora of competing values and ethical issues connected to the actual and potential influences of technology. This includes philosophical, historical, economic, legalistic, cultural and political perspectives on technological systems.

Philosophical perspectives provides a powerful knowledge building tool for entrenching Technoethics within a larger discourse and tradition. For instance, Mitcham's (1994). *Thinking through Technology: The Path between Engineering and Philosophy* explores changing conceptualizations of "technology" from early Greek to modern civilization. This text defines technique as "the totality of methods rationally arrived at and having absolute efficiency (for a given stage of development) in every field of human activity" (308). The text explores various conceptions of "technology" (technology as object, technology as knowledge, technology as activity, technology as volition, etc.) in an effort to expand understanding of technology to help inform current work. It also provided a focused philosophical inquiry into a specific area of human activity (Engineering) that could be replicated in other domains relying heavily on science and technology. Other examples of philosophical inquiry into focused areas of human activity include Johnson's (1994) *Computer Ethics*. Philosophical perspectives on general and focused areas of technology are useful knowledge building tools for technoethical inquiry.

Historical perspectives in technoethical inquiry are helpful in revealing underlying influences directly affecting science and technology innovation (micro level) as well as providing insights into general historical contexts to help situate technoethical inquiry (macro level). For an example of micro level studies, Cutcliffe's (2000) *Ideas, Machines, and Values: An Introduction to Science, Technology, and Society Studies* examines the specific roots of STS from the 1960s to the end of the 1990s. Cutcliffe remarked, "When we try to pick out anything by itself, we find it hitched to everything else in the universe" (vii). This historical inquiry into the state of STS offers valuable insights into the historical context that affects the meaning and use of technology in society. In terms of macro level historical studies, Pacey's (1990) *Technology in World Civilization: A Thousand Year History* provides an excellent review of technological development in both Eastern and Western Civilization beginning in 1000 AD in China. The study highlights major technological breakthroughs transferred from the Eastern to the Western world (hydraulics, iron production, shipbuilding, textiles, printing, gunpowder) and from the Western to the Eastern world (factory production, scientific principles of knowledge organization). This historical research was extremely important in demonstrating how technology transfer between Eastern and Western civilization was connected to technology transfer strategies, cultural influences, and the ability of societies to create technological dialogues about local experiences with technology. Both micro and macro level historical analyses are useful knowledge building tools for technoethical inquiry when technological phenomena has to be contextualized and situated within society.

Economic perspectives are also powerful knowledge building tools in technological inquiry and have been used to address key questions like, "How are technological advances introduced into economic activity and what underlying processes are important to examine?" Rosenberg's (1982)*Inside the Black Box: Technology and Economics* is one influential work that examined the nature of technology progress and its impact on production from an economic standpoint. It demonstrated how new technologies influence key factors of interest to economists including, the diffusion of new technologies, technology transfer, government policy influences, market demand influences, commercial innovations, and industry relationships. The text identified a number of problems associated with technological progress including the high cost of technology development, low predictability concerning long term performance outcomes of new technologies, and time needed for technologies to be tested by users. He stated, "In an economy with complex new technologies, there are essential aspects of learning that are a function not of the experience involved in producing the product but of its utilization by the final user (122). Another useful economic perspective derives from Grübler's (2003)*Technology and Global Change*, which explores from an economic standpoint how technology has influenced change in the society and the environment over the last 200 years. The text reviews negative impacts of technology (urban pollution, smog, global warming) in relation to global environmental change within a capitalistic marketplace to help inform long-term policy planning in industry and government. He states, "The fundamental impulse that acts and keeps the capitalistic engine in motion comes from the new consumers' goods, the new methods of production or transportation, the new markets, the new forms of industrial organization that capitalist enterprise creates" (41).

Political perspectives can provide key knowledge into public perception and power structures that often influence technological innovation. Salomon's(1973) *Science and Politics* delved into the evolution of science and technology as a consolidation of national policies and public institutions. Rather than linking science and technology policy creation to advances within, this text illustrates the shaping power of political institutions and their organization. In a slightly different vein, Bell's (1973)*The Coming of Post-Industrial Society: A Venture in Social Forecasting. New York: Basic Books* investigated the rise of a post-industrial society driven by an information-led and service-oriented economy that would supplant the industrial society as the dominant system or political power. This text traces a societal shift from manufacturing to service industry and the rise of a knowledge based economy driven by science, technology and technical elites. Another political study that has added depth to technoethical inquiry is Sarewitz's (1996)*Frontiers of Illusion: Science, Technology, and the Politics of Progress*. This text focuses specifically on diminishing federal funding for research in the US due to research accountability issues. From a political perspective Sarewitz argues that researchers should invest more effort to address societal concerns about the dangers of new technologies and scientific advances to ensure societal support of continued research and to enhance chances for human survival into the long-term future. The text argues, "New products, new industries, and more jobs require continuous additions to knowledge of the laws of nature, and the application of that knowledge to practical purposes" (17). Political perspectives help uncover new knowledge about key catalysts from outside science and technology that that help drive scientific research and policy in society.

Cultural and critical perspectives help provide technoethical inquiry useful knowledge about underlying cultural meaning and power structures that often remain implicit and difficult to explain. An exemplar work in this area is Tichi's (1987)*Shifting Gears: Technology, Literature, Culture in Modernist America*. Tichi (1987) explores how American technology between the 1890s and 1920s influenced American culture through the language used by writers of the time (e.g., Ernest Hemingway and John Dos Passos). Ticcchi argued that technology was entrenched in the culture and the minds of poets and novelists in society. She noted, "There's nothing sentimental about a machine, and: A poem is a small (or large) machine made of words" (270). In a similar vein Yaszek (2002) examines how post WWII America described the potential effects of new information and technologies on life and society through science fiction. The book delves into science fiction work (*Neuromancer, Snow Crash, Blade Runner, Cyberpunk*) that have linked human activity to technologies in creative ways to demonstrate how this technology could alter life and society. Based on a critical theory perspective, Feenberg (1991) draws on leading socialist theory (i.e., Marx) and contemporary scholars (i.e., Habermas, Ellul, Foucault) to examine the social use of technology in society. Cultural and critical perspectives contribute unique knowledge that can be used help technoethical inquiry better serve the needs of various cultures and social groups within society.

Legalistic perspectives and technology oriented legislation in areas such as cyber law (Trout, 2007) help provide technoethical inquiry useful knowledge about normative perspectives and policies that help regulate technology and shape how technology advances. For instance, to combat cyber-bullying, the U.S. Assembly Bill 86 2008 was introduced to provide school administrators the authority to discipline students for bullying others offline or online (Surdin, 2008). A number of US State privacy laws have been created to protect the rights of citizens against violations of human rights to privacy (see BBBBOnline, 2008). Other legalistic document such as the Canadian Copyright Act (Department of Justice Canada, 2006) and Digital

Millennium Copyright Act (Digital Millennium Copyright Act, 1998) are intended to protect author rights and regulate the appropriate use of authored works. Legalistic perspectives are often found in legal cases as seen as illustrated by this excerpt from the Internet Library of Law and Court Decisions (2009 John Doe v. SexSearch.com 502 F.Supp. 2d 719, Case No. 3:07 CV 604 (N.D. Ohio, August 22, 2007)

Court holds that the Communications Decency Act ("CDA"), 47 U.S.C. Section 230, immunizes operator of online adult dating service from claims arising out of a user's false statement in her user-profile that she was over 18. Relying on this profile, plaintiff met and had consensual sexual relations with a minor, for which he was subsequently arrested. Plaintiff brought this suit, seeking redress. Importantly, the contract between the parties expressly provided that SexSearch.com does not "assume any responsibility for verifying the accuracy of the information provided by other users of the Service." Because plaintiff sought to hold SexSearch.com, a provider of Interactive Computer Services, liable for its publication of content authored by another, his claims, whether couched as breach of contract, fraud or negligent misrepresentation, were barred by application of the CDA. Plaintiff's breach of contract claim similarly failed because SexSearch did not assume responsibility for verifying the age of users.

The use of such of legalistic frameworks to address new technology developments (i.e., cyberlaw, neuroloaw, nanolaw) provide valuable insight into appropriate and inappropriate technology use for enhancing technoethical inquiry in areas of technology development where legal concerns are applicable.

The abovementioned perspectives illustrate the role of multiple types of expertise (and experts) that can be mobilized to inform technoethical inquiry. Historians may contribute valuable insights into prior work relevant to current development and social and ethical implications. Scientists may focus on advancing technological research and development while attending to codes of responsible conduct and professional responsibilities. Ethicists and ethical theorists may contribute insights through ethical theories focused on what social and ethical elements should be upheld. Technoethical inquiry may involve the integration of many perspectives depending on the level of technological system relations within the context of deployment. From this, it can be seen that a proper technoethical inquiry cannot be limited to any one discipline. Rather, it strives to integrate multiple perspectives derived from various disciplinary knowledge bases in order to provide a more balanced view of the complexity, specificity, and scope of technological development in the world.

IDENTIFYING KEY AREAS AND ISSUES IN TECHNOETHICS

Advancing science and technology in diverse areas of human activity within the evolving knowledge society is giving rise to equally diverse set of ethical issues and questions. What are the responsibilities of technologists to those affected by their work? How should responsible be assigned for the negative impacts of Engineering on society? What are the ethical responsibilities of Internet researchers to research participants? How do advances in educational technology affect access to new educational resources and the growing digital divide? Who should have ownership and control of harvested DNA, human tissue and other genetic material? How can virtual organizations resolve communication conflicts and satisfy stakeholder interests? What are the ethical responsibilities of professionals using technology to contribute to helping mankind ? How do we assign responsibility in environmental construction and management? What are the potential risks with nanotechnology applications and how should responsibility be assigned? Who should be responsible for controlling advanced military technology?

As will be elaborated in subsequent chapters, Technoethics is a field of study and research constituted by core branches focusing key areas of human activity affected by technology. Figure 2 illustrates core branches in Technoethics followed by a brief introduction to each branch. This breakdown is a modified version of core branches identified by myself in the *Handbook of Research of Technoethics* (Luppicini, 2009).

Technoethics and Cognition is a branch of Technoethics concerned with technological processes within human mind, artificial agents, and society. This includes a variety of topics including artificial morality, moral agents, technoethical systems, technoethical mind, techno-addiction and ethical intervention. In line with topics discussed earlier in the book (see chapter 3), scholarship within philosophy technology and cognitive sciences from the 1960s onward have introduced new concepts (collective intelligence, distributed cognition, artificial intelligence, artificial moral agents) to examine questions of agency and ethical responsibility in humans and non-human entities. Although humans are considered autonomous agents with morals, the question arises as to whether artificial autonomous agents (artificial entities which resemble human autonomous agents in their ability to initiate events and processes) may be conceptualized as moral agents in so far as morality is reflected in their functional abilities. A leader in research on artificial agency in Technoethics, Sullins (2009) argues:

Artificial agents created or synthesized by technologies such as artificial life (ALife), artificial intelligence (AI), and in robotics present unique challenges to the traditional notion of moral agency and that any successful technoethics must seriously

Figure 2: Branches of Technoethics

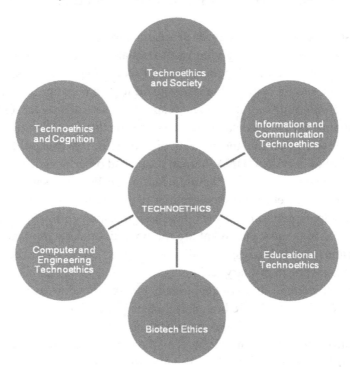

consider that these artificial agents may indeed be artificial moral agents (AMA), worthy of moral concern (205).

This raises a number of questions for Technoethics and cognition: Is moral agency limited to human agency? If artificial agency was found to be functionally similar to humans agency should these artificial entities be awarded moral rights and moral responsibilities similar to that of human agents? Is it possible it is possible for us to create artificial moral agents? If so, under what conditions and with what aims?

Other areas of research in this area explore ethical dilemmas of technology and psychological effects such as technology over dependency, addiction and withdrawal. For instance, Cortes (2009) argues that ICT overuse led to a number of negative consequences including psychological addition, feelings of isolation, passivity, and emotional stress. What constitutes technology abuse and how can it be avoided? How does technology dependency negatively affect the mental health of humans and how can this be avoided? What ethical considerations need to be made when deploying technologies which can alter human consciousness, sense of self, and sense of others?

Technoethics and Society is a branch of Technoethics concerned with the ethical use of technology to promote the aims of contemporary society. This includes a variety of topics and areas including digital property ethics, technoethics and social theory, technoethics and law, technoethics and science, technoethics and art, organizational technoethics, and global technoethics. Because of the centrality of information and communication technology in the world economy, digital property ethics are an important area of societal concern within social theory and legal discourse. Art and science, as social institutions, have a real impact and technoethical significance within the evolving knowledge society that requires careful consideartion. Organizational technoethics focuses on how technological advances (i.e., ICT's, distributed scientific collaborations, new organizational structures) are redefining organisations and how they operate within an evolving knowledge economy. The shift of organizations to greater knowledge creation combined with the integration of technological advances has led to changing working relations (i.e., telework, videoconferencing, shareware applications, virtual organizations, virtual organizational alliances) and a number of ethical dilemmas in society. Global technoethics extends the analysis of technology and ethics to a global scale dealing with questions such as technology and developing nations, cultural assimilation and exploitation of foreign populations and resources.

Computer and Engineering Technoethics represents a branch of Technoethics concerned with the ethical aspects of computer technology and engineering in contemporary life and society. Key topic areas of include computer environmental technoethics, military technoethics, nanoethics, nuclear ethics, etc. This branch of Tehnoethics draws on work in computer ethics and the concern with human use of computer technology in a number of areas (i.e., computing in the workplace, graphic interfaces, visual technology, artificial intelligence, and robotics). Johnson (1985) defined Computer Ethics as a new form of ethical inquiry influenced by the advancement of computer technology which "pose new versions of standard moral problems and moral dilemmas, exacerbating the old problems, and forcing us to apply ordinary moral norms in uncharted realms" (p.1). This branch also draws on engineering ethics and public concern over large scale research and innovation (I.,e., military operations and nanotechnology research). It focuses on the conduct of engineers and their moral responsibilities to the public in areas where new scientific and technological innovations give rise to conflicting societal needs as reflected in public and media. Key areas of technoethical concern include Aerospace, Agriculture, Architectural, Bioengineering, Chemical Engineering, Civil Engineering, Electrical Engineering, Industrial Engineering, Mechanical Engineering, and nuclear energy creation. Johnson's (1991) *Ethical Issues in Engineering* and Unger's (1982) *Controlling Technology: Ethics and the Responsible Engineer* are seminal publications in this area. One major area of focus, environmental technoethics deals with ethical aspects of technological innovations that impact the environment and life including transport, urban planning, mining, sanitation, marine contamination, and terraforming (altering the environmental conditions of another planet to make it habitable by humans). Another major area of focus, military technoethics, deals with ethical issues associated with technology use in military action. It focuses attention on understanding war-technology and how it redefines relationships. This is particularly important regarding the appropriate use of controversial new technologies (i.e., biological warfare, nuclear weapons) in situations where the military operates. Hartle's (1989)*Moral Issues in Military Decision Making* is an example of technoethical inquiry into ethical issues governing military decision from the position of an American army scholar. A rising area of technoethical scholarship under the heading of nanoethics, focuses on ethical issues associated with developments in nanotechnology (the alteration of matter at the level of atoms and molecules). Nanotechnology has contributed to the advancement of research in a variety of disciplines including, Computer Science, Engineering, Biology, and Chemistry (Drexler, 1986) while raising health and safety issues and public debate (Hunt & Mehta, 2006). Hunt and Mehta's (2006)*Nanotechnology: Risk, Ethics and Law* reviews key developments in nanotechnology risks and ethical issues connected to this potentially dangerous technological innovation.

Information and Communication Technoethics represents a branch of Technoethics concerned with the ethical aspects connected to information and communication technology. Key topics include cyberethics, cyber pornography, cybercrime, cyber-stalking, internet ethics, media ethics, netiquette, etc.). Theoretical work in Information Ethics is becoming increasing applied in Computer Ethics (and other areas) in identifying processes and interactions within the informational environment (Floridi & Sanders, 2003). In expanding the scope of Computer Ethics, Floridi and Sanders (2003) argued that information is "a necessary prerequisite for any morally responsible action to being its primary object." This was based on Floridi and Sander's notion of "infosphere", which described the informational environment constituted by informational entities, including processes and interactions (Floridi & Sanders, 2003).Other core characteristics of this branch include communication processes and ethical issues connected to mass media and communication technology. This stems from scholarly work in communication media and media discourse connecting communication and technology considerations. McLuhan's *The Gutenberg Galaxy: The Making of Typographic Man* (1962), *Understanding Media: The Extensions of Man* (1964) and Habermas' (1992) Discourse Ethics are important works in this domain. McLuhan (1962) demonstrated how various communication technologies influence cognitive organization and McLuhan (1964) linked specific media characteristics to societal affects and demonstrated the need for human awareness of technology's cognitive effects in shaping individual and societal conceptions. Habermas' (1992) discourse ethics examined communicative structures and explained the obligatory nature of morality rooted in communicative rationality. This led to other approaches to discourse ethics and online deliberation (Benhabib, 1992; Thorseth, 2006) that helped to ground ethical inquiry within processes of communication in a variety of communication contexts including technical communications (Allen, & Voss, 1997), virtual organizations (Rotenberg, 1998) and online communication (Mann & Stewart, 2000).

Internet Ethics and Cyberethics are key areas of Information and Communication Technoethics that deal with new challenges revolving around the growth of the Internet and new computing technologies for the Internet (e.g., spyware, antivirus software, web browser cookies). It also deals with ethical dilemmas such as, unauthorized access, threatening the integrity of computer-based information, and threatening the privacy of users. Cyberethics extends Internet Ethics into considerations of the Internet as a democratic technology for guiding digital citizenship and responsible behavior in technology use (cyber democracy). It also deals with ethical dilemmas such as cyber stalking, Internet pornography, cyber identity theft, identity fraud, and using the Internet to 'fish' for credit card numbers, bank account information and other personal information to be used illegally (phishing).

Educational Technoethics is the branch of Technoethics dealing with ethical issues associated with using technology for educational aims (Cortés, 2005). Key topics include cyber-bullying, cyber democracy, digital divide, e-learning ethics, emancipatory educational technology, professional technoethics, technoethical assessment and evaluation Scholars from areas of Education and Educational Technology like Nichols (1987, 1994) and Boyd (1991) examine negative aspects of educational technology and highlight emancipatory uses of educational technology as a means to leverage ethical guidelines and values concerning the use of new technology. It also deals with the misuse of educational technology such as in the case of plagiarism (Lathrop & Foss, 2000; Underwood & Szabo, 2003) and questionable issues connected to detection tools to deter internet plagiarism, cyber-bullying, and professional technoethics. Cyber-bullying involves the use of information and communication technologies to support deliberate, repeated, and hostile behavior by an individual or group, that is intended to harm others. Professional technoethics deals with all ethical issues connected to the use of technology within professional conduct. It delves into the professional training and ethical responsibilities within professions that actively use new technologies (e.g., Engineering, Journalism, Medicine, etc) and examines different codes of conduct and policies for professionals working with technology in different areas of human activity. For instance, computer scientists have ethical responsibilities in designing, developing, and maintaining computer hardware and software systems while journalists have ethical responsibilities concerning the protection of informant identity, accessing information, and disseminating information for public viewing.

Biotech Ethics is a branch of Technoethics dealing with ethical conflicts connected to the spread of biotechnologies in medical research, health care, and industrial applications (e.g., abortion, animal rights, eugenics, euthanasia, *in vitro* fertilization, and reproductive rights). Key topics include cloning ethics, e-health ethics, telemedicine ethics, medical, research ethics, genetic ethics, neuroethics, and sport and nutrition technoethics. This branch draws on Jonas (1985) seminal work examining social and ethical problems in and the Human Genome Project by the U.S. Department of Energy in 1986 sparked raised serious ethical issues arising in many areas of biological technology including, cloning, gene therapy, genetically modified food, genetic engineering, drug production, and stem cell research. Innovation in biotechnology leverages work in many areas (e.g., genetics, biochemistry, chemical engineering, information technology, biometrics, tec.) while creating new ethical questions to deal with (Schummer & Baird, 2006).

Table 1 presents the main areas of Technoethics, along with selected issues and guiding questions. For better organization, some branches of Technoethics are combined in the chapters where logical pairing was possible. This is intended to convey the close relationship between areas treated separately in this chapter.

DISCUSSION QUESTIONS

- How should the public be informed of engineering risks?
- Who should be responsible for controlling advanced military technology?
- What would be the consequences if artificial entities were treated as autonomous and responsible moral agents?
- How do we deal with Internet abuse and misuse such as piracy, pornography, and hate speech?
- Who should have ownership and control of harvested DNA, human tissue and other genetic material?

DISCUSSION

This chapter grounds Technoethics in a social systems approach guiding technoethical inquiry. It describes how technoethical inquiry seeks to understand technology as a complex and multifaceted system defined by sets of technological-human relations. This chapter helped address the variety of ways that technology manifests itself within society in terms of multiple technological levels and multiple-

Table 1. Selected issues and guiding questions in technoethics

Branches of Technoethics	Selected Issues	Selected Guiding Questions
Computer and Engineering Technoethics	professional codes of ethics, computer ethics, environmental technoethics, military technoethics, nanoethics, nuclear ethics, etc.	How do we gauge the ethical use of new computer technologies? What are the ethical considerations in computer production? Who should be responsible for the negative impacts of Engineering on society? How should the public be informed of engineering risks? How do we deal with the fact that many Engineering applications can be used for evil? How do we help improve the lives of people living in cities suffering from infrastructural decay? What are the rights of individuals in environmental construction and management? What are the potential health and safety risks with nanotechnology applications and who is responsible? What are the rights of individuals in affecting nanotechnology applications? What are the actual and possible risks with nuclear weapons? Who should be responsible for controlling advanced military technology?
Technoethics and cognition	artificial morality, ethical agents, technoethical systems, technoethical mind, techno-addiction and ethical intervention, etc.	What would be the consequences if artificial entities were treated as autonomous and responsible moral agents? What rights and responsibilities should they be granted? Are the conditions for moral responsibility the same as for legal responsibility or different? Should an artificial entity be held legally liable? If so, under which conditions? Is personhood a necessary prerequisite for ascribing moral agency or could moral agency be attributed to artificial entities?
Information and Communication Technoethics	cyberethics, cyber pornography, cyberterrorism, cybercrime, cyberstalking, internet ethics, media ethics, netiquette, etc.	How do we deal with Internet abuse and misuse such as piracy, pornography, and hate speech? Who should have access to the Internet and who should be in control? How should we protect young Internet users from unnecessary risks derived from Internet use? What are the ethical responsibilities of Internet researchers to research participants? What are the ethical responsibilities of Internet researchers to protect the identity and confidentiality of data derived from the Internet? Is democratic decision making possible through online deliberation?
Educational Technoethics	cyber-bullying, cyber democracy, digital divide, e-learning ethics, emancipatory educational technology, professional technoethics, technoethical assessment and evaluation, etc.	How do advances in educational technology affect access to new educational resources and the growing digital divide? How do new educational technologies influence multicultural education and inclusion? What are the ethical considerations in online plagiarism and how should such situations be dealt with How do advances in educational technology affect access to new educational resources and the growing digital divide? How will each technological innovation affect people in the short term and the long-term? What are the ethical responsibilities of professionals using technology to contribute to helping mankind ? Do the outcomes of professional work with technology prioritize the rights of some individuals over others?

continued on following page

Table 1. continued

Branches of Technoethics	Selected Issues	Selected Guiding Questions
Technoethics and Society	technoethics and social theory, technoethics and law, organizational technoethics, global technoethics, e-business ethics, technoethics and knowledge management, technoethics and work, etc.	What are the ethical implications of allowing monopolies to control software and hardware development? What are the responsibilities to stakeholders for those involved in computer technology management? How can virtual organizations resolve communication conflicts and satisfy stakeholder interests? How can virtual organizations resolve communication conflicts and satisfy stakeholder interests? What are the ethical considerations connected to virtual organizational alliances? What are the ethical dilemmas that arise when developed nations conduct research in developing nations? What are the ethical implications of outsourcing work operations to other areas of the globe?

perspectives relevant to the study of technology as a multi-faceted social system. The chapter also provided a breakdown of core branches of Technoethics, namely, Computer and Engineering Technoethics, Technoethics and cognition, Information and Communication Technoethics, Educational Technoethics, Biotech Ethics, and Technoethics and Society.

REFERENCES

Allen, L., & Voss, D. (1997). *Ethics in technical communication: Shades of gray.* New York: Wiley Computer Publishing, John Wiley & Sons, Inc.

BBBBOnLine. (2008). *A Review of Federal and State Privacy Laws.* BBBBOnLine, Inc. and the Council of Better Business Bureaus, Inc. Retrieved June 9, 2009, from http://www.bbbonline.org/Understanding-Privacy/library/fed_statePrivLaws.pdf

Bell, D. (1973). *The coming of post-industrial society: A venture in social forecasting.* New York: Basic Books.

Benhabib, S. (1992). *Situating the self.* Cambridge, UK: Polity Press.

Boyd, G. (1991). The shaping of educational technology by cultural politics, and vice versa. *Educational and Training Technology International, 28*(3), 87–96.

Bunge, M. (1977). Towards a technoethics . *The Monist, 60*(1), 96–107.

Cortés, P. A. (2005). Educational technology as a means to an end. *Educational Technology Review, 13*(1), 73–90.

Cutcliffe, S. (2000) *Ideas, machines, and values: An introduction to science, technology, and society studies.* Lanham, MA: Rowman and Littlefield

Danielson, P. (1992). *Artificial morality: Virtuous robots for virtual games.* London: Routledge.

Department of Justice Canada. (2006). *Copyright Act, R.S., 1985, c. C-42.* Retrieved June 1, 2009 from http://laws.justice.gc.ca/en/C-42/index.html

Digital Millennium Copyright Act. (1998). Retrieved June 3, 2009, from http://searchcio.techtarget.com/sDefinition/0,290660,sid19_ gci904632,00.html

Drexler, E. (1986). *Engines of creation: The coming era of nanotechnology.* New York: Anchor Press/Doubleday.

Feenberg, A. (1991). *Critical theory of technology.* New York: Oxford University Press.

Floridi, L., & Sanders, J. (2003). Computer ethics: Mapping the foundationalist debate. *Ethics and Information Technology, 4*(1), 1–24. doi:10.1023/A:1015209807065

Grübler, A. (2003). *Technology and global change.* Cambridge, UK: Cambridge University Press.

Habermas, J. (1992). *Moral consciousness and communicative ethics.* Cambridge, MA: MIT Press.

Hartle, A. (1989). *Moral issues in military decision making.* Lawrence, KS: University of Kansas Press.

Hinduja, S., & Patchin, J. W. (2009). *Bullying beyond the Schoolyard: Preventing and Responding to Cyberbullying.* Thousand Oaks, CA: Sage Publications.

Hunt, G., & Mehta, M. (2006). *Nanotechnology: Risk, ethics and law.* London: Earthscan Book.

Internet Library of Law and Court Decisions. (2009). *John Doe v. SexSearch.com.* Retrieved June 1, 2009, from http://www.internetlibrary.com/internetlib_subject.cfm?TopicID=1

Johnson, D. (1985) *Computer ethics.* Englewood Cliffs, NJ: Prentice-Hall.

Johnson, D. (1991), *Ethical issues in engineering.* Englewood Cliffs, NJ: Prentice-Hall.

Jonas, H. (1984). *The imperative of responsibility, in search of an ethics for the technological age.* Chicago: University of Chicago Press.

Lathrop, A., & Foss, K. (2000). *Student Cheating and Plagiarism in the Internet Era: A wake-up call.* Englewood, NJ: Libraries Unlimited Inc.

Mann, C., & Stewart, F. (2000). *Internet communication and qualitative research: A handbook for researching online.* London: Sage.

Marcuse, H. (1964). *One dimensional man* Boston: Beacon Press

McLuhan, M. (1962). *The gutenberg galaxy.* Toronto: McGraw Hill.

McLuhan, M. (1964). *Understanding media: The extensions of man.* Toronto: McGraw Hill.

McLuhan, M., & Parker, H. (1969). *Counterblast.* New York: Harcourt, Brace, & World.

Mitcham, C. (1994). *Thinking through technology: The path between Engineering and Philosophy.* Chicago: University of Chicago Press.

Moor, J. H. (1985). What is computer ethics. In T. W. Bynum, (Eds.). *Computers and ethics* (pp. 266-275).New York: Basil Blackwell.

Moor, J. H. (2005). Why we need better Ethics for emerging technologies. *Ethics and Information Technology, 7,* 111–119. doi:10.1007/s10676-006-0008-0

Mumford, L. (1934). *Technics and civilization.* New York: Harcourt Brace.

Nichols, R. G. (1987). Toward a conscience: Negative aspect of educational technology. *Journal of Visual/Verbal Languaging, 7*(1), 121-137.

Nichols, R. G. (1994). Searching for moral guidance about educational technology. *Educational Technology, 34*(2), 40–48.

Pacey, A. (1983). *The culture of technology.* Cambridge, MA: MIT Press.

Polanyi, M. (1966). *The tacit dimension.* New York: Doubleday & Co.

Rosenberg, N. (1982). *Inside the black box: Technology and economics.* New York: Cambridge University Press.

Rotenberg, M. (1998). Communications privacy: implications for network design. In Stichler, R. N. & Hauptman, R. (Eds.). *Ethics, Information and Technology Readings.* Jefferson, NC: McFarland & Company, Inc., Publishers.

Salomon, J. (1973). *Science and politics.* Cambridge, MA: MIT Press.

Sarewitz, D. (1996). *Frontiers of illusion: science, technology, and the politics of progress.* Philadelphia: Temple University Press.

Schummer, J., & Baird, D. (2006). *Nanotechnology challenges: Implications for philosophy, ethics and society.* London: World Scientific.

Surdin, A. (2008). States Passing Laws to Combat Cyber-Bullying. *Washingtonpost. com.* Retrieved June 1, 2009, from http://www.washingtonpost.com/wpdyn/content/article/2008/12/31/AR2008123103067.html.

Thorseth, M. (2006). Worldwide deliberation and public reason online. *Ethics and Information Technology, 8,* 243–252. doi:10.1007/s10676-006-9116-0

Tichi, C. (1987). *Shifting gears: Technology, literature, culture in modernist america.* Chapel Hill, NC: University of North Carolina Press.

Trout, B. (2007). *Cyber law: A legal arsenal for online business.* New York: World Audience, Inc.

Underwood, J., & Szabo, A. (2003). Academic offences and e-learning: Individual propensities in cheating. *British Journal of Educational Technology, 34*(4), 467–477. doi:10.1111/1467-8535.00343

Unger, S. (1982). *Controlling technology: Ethics and the responsible engineer.* New York: Holt, Rinehart and Winston.

Vallenella, E. (1999). From meta-technology to ecology. *Bulletin of Science, Technology & Society, 19*(5), 411–415. doi:10.1177/027046769901900509

Wiener, N. (1954). *Human use of human beings.* New York: Houghton Mifflin.

Yaszek, L. (2002). *The self wired: Technology and subjectivity in contemporary narrative.* New York: Routledge

ADDITIONAL READING

Cavalier, R. (2005). *The impact of the Internet on our moral lives.* NY: State University of New York Press.

Orwell, G. (1949). *1984.* Harcourt Brace Jovanovich, Inc.

Passmore, J. (1974). *Man's responsibility for nature.* London: Duckworth.

Reich, C. (1970). *The greening of America.* New York: Random House

Rolston, H. (1975). Is there an ecological ethic? *Ethics, 85,* 93–109. doi:10.1086/291944

Rousseau, J. J. (1762). The social contract. New York: Hafner Publishing Company.

Schmidtz, D., & Willott, E. (*2002*). *Environmental ethics: What really matters, what really works.* New York: Oxford University Press.

Stone, C. D. (1972). Should trees have standing? *Southern California Law Review*, *45*, 450–501.

Sullins, J. P. (2006 a). Ethics and artificial life: From modeling to moral agents. *Ethics and Information Technology*, *7*, 139–148. doi:10.1007/s10676-006-0003-5

Sullins, J. P. (2006 b). When is a robot a moral agent? *International Review of Information Ethics*, *6*, 23–30.

Section 3
Technoethics and Its Deployment

Chapter 6
Information and Communication Technoethics

INTRODUCTION

Where is the life we have lost in living? Where is the wisdom we have lost in knowl-edge? Where is the knowledge we have lost in information?" -Vanderburg, 2005, p. 387.

Vanderburg's (2005) *Living in the Labyrinth of Technology*, describes the seemingly ambivalent state of life and meaning within a technological society. The ubiquity and invisibility of advancing information and communication technologies (ICT's) challenges individuals sense of self and society, and their understanding of how meaning is communicated, by whom, for what purpose, and with what outcomes.

DOI: 10.4018/978-1-60566-952-6.ch006

The convergence of information, communication, and technology has become an important concern in academia as is apparent in the intersecting interests of technology studies, information studies, and communication studies in areas related to the role of technology in social interaction, meaning creation, identity formation, culture, and information exchange. This intersection of fields is partly due to the convergence of information and communications with advancing technological innovation. This has given rise to the ever-expanding convergence in academic research within communications and technology studies. This is exemplified through an amassing body of research publications focusing on technology, information, and communication, along with continued growth of technology and communication oriented research activities carried out within professional associations (Society for Social Studies of Science [4S] and the European Association for the Study of Science and Technology [EASST], International Communication Association, Canadian Communication Association).

The study of social and ethical dimensions of information and communication technologies (ICT's) is among the most fruitful areas of interdisciplinary scholarship emerging at the intersection of communication with other fields dealing with technology, society, and human life. This is partly in response to the public demand for academic work that delves into underlying developments, ethical dilemmas, and real-life controversies surrounding the web of new relations created by new communication technologies. This has reinforced knowledge sharing and the cross-fertilization of related fields by encouraging the exchange of ideas and expertise pertaining to the ethical studies of communication and technology. It has also provided new conceptual grounding for previously programs of ethical inquiry pursued within niche areas of technology studies and communication studies (i.e., media ethics, cyber ethics, ethical perspectives in science and technology studies, information ethics). The resulting influence of this interdisciplinary scholarship has helped substantiate communication technoethics as a promising area of interdisciplinary scholarship.

Technoethics in our contemporary knowledge society is marked by powerful information and communication technologies that shape and reshape communication, culture, politics, and the economy. This places Information and Communication Technoethics as a central branch of technoethical inquiry. The discussion that follows deals with the emergence of Information and Communication Technoethics as a branch of Technoethics that focuses on ethical and social aspects of advancing information and communication technologies in all areas of life and society. It begins with a background sketch of Information and Communication Technoethics rooted in technological developments in contemporary society, along with key academic developments within communication studies, technology studies, interdisciplinary studies, and philosophy. Then, it reviews current work on ethical aspects of ICT's that increasingly mediate important aspects of how individuals live and interact. The discussion that follows attempts to raise understanding about how technological progress introduces new social and ethical challenges in work and life.

BACKGROUND

Change and Development in Information, Communication and Technology

Major upheavals brought on by new technological developments, changes in how people exchange information and communication, and the nurturing in of new economic possibilities from the 1960s to present day have not only shaped the direction of technology and society, but have also shaped academia. This can be attributed to a number of factors including technological changes, the convergence of technology studies, information studies, and communication studies, and the rise of social and ethical critiques in response to controversial developments and negative consequences connected to technology, information exchange, and the knowledge economy. The result of these developments have created the context for interdisciplinary research on technology and ethics.

Technological Change

Perhaps the most important technological development in the 20th century is the Internet. The Internet got its start in 1969 when a precursor to the Internet called the ARPANET (Advanced Research Projects Agency Network) was created by the United States Department of Defense to serve as a networking system for information storage and exchange. Despite the practical benefits, widespread public use of the Internet began to emerge in the late 1980s and early 1990s as computer technology evolved and became more affordable. This changed the face of information exchange and communications around the world. Concurrent with the rise of Internet popularity also came a variety of new social and ethical challenges to deal with. For instance, in the area of policy development, the Internet Architecture Board (IAB) created a policy concerning Internet Ethics in 1989 to help to guard against unethical Internet use, such as gaining unauthorized access to Internet resources and jeopardizing the privacy and rights of users. The Internet Architecture Board (IAB) considered the following as unethical activities:

- Seeking to gain unauthorized access to the resources of the Internet
- Disrupting the intended use of the Internet

- Wasting resources (i.e., people, capacity, computer)
- Destroying the integrity of computer-based information
- Compromising the privacy of users (Internet Architecture Board, 1989)

From the 1990s to today, the popularization of the Internet, along with the development of powerful information and communication technologies (ICT's) began playing a pivotal role in transforming the nature of experience and communicating with others (and ourselves) across many areas of life and society (e.g., education, healthcare, work, and relationships, and community). This has changed the face of communications in regards to human agency, research practices, knowledge creation, and information sharing. The shift to ICT convergence has triggered a shift in human agency from a low to medium level of technological mediation to a high level of technological mediation. For instance, the advent of tele-work and virtual organizations allow work communications to be conducted partially or entirely while separated from co-workers and managers. A similar separation has occurred in the use of social networking software (e.g., Facebook, friendster, orkut, my space, etc.) to maintain personal relations and social interaction. This, in turn, has given rise to a number of new questions concerning where humans fit within a technological landscape and vice versa. How do individuals communicate and exchange information in a world of converging ICT's? Could high degrees of ICT convergence eliminate the need for humans to think, make decisions, or control their lives? How does increased technological mediation affect human agency?

In response to the growing intertwining of technology in work and life, technological pessimists have argued that technology threatens to replace humans and their role in society. However, this is not the case in most areas. Increased use of ICT for knowledge creation has not reduced the importance of human agency. Rather, it has shifted the role of human agency. For instance, Hakken's (2003)*Knowledge Landscapes of Cyberspace* provided a social informatics analysis of the role of human agency in knowledge creation within computer mediated environments. The text offers an insightful look at the current characteristics of informational technologies, how they are used, and how they are influenced by individual interests and power relations. In this sense, the importance of human agency has not changed. What has changed are the tools for communication and how human agency is understood. It is now possible to automate how communications are sent and received as is the case in pre-planning database or weblog updates while one is away. This can give the impression of human presence within a dynamic communication system when this is not the case. This does, however, create new challenges in understanding human agency in the face of technological mediation where human agency is one level removed from the communication context. As will be shown later in this discussion, this raises new questions concerning ethical conduct and social responsibility within technological mediated environments that expands the scope of the field of communications in new ways.

Convergences In Academic Work on Technology, Information and Communication

The abovementioned technological transformations in life and society are mirrored in academic scholarship. The fields of information science, communication, and technology studies (which deals with the role of technology in life and society) have been at the nexus of the these important transformations. The scientific approach to the study of information (information science) provided much of the groundwork needed for Information and Communication Technoethics. The importance of information and communication was first highlighted during the information revolution pioneered in the 1940s and 1950s through seminal works including Shannon and Weaver's (1949)*The Mathematical Theory of Communication* and Wiener's (1948) Cybernetics. The outgrowth of this led to the popularization of information science as an interdisciplinary area of research focused on the collection, classification, management, and dissemination of information within various contexts. Work within information science has roots in numerous existing disciplines and fields (e.g., computer science, cognitive science, commerce, communications, law, library science, mathematics, social sciences, etc.) and an orientation towards the study of information and communication technologies (ICT's) and information systems. These developments have made information science and communications a vital and influential area of research in the 21st century. This can be attributed to a number of factors including the convergence of technology studies and communication studies, core work in media studies, and the rise of social and ethical critiques in response to controversial developments and negative consequences connected to technology, communication, and the knowledge economy.

First, the convergence of technology studies, information studies (information science), and communication studies has expanded the scope and potential impact of communication research and practice. Also, work from a variety of pioneering scholars of technology and communications in the 1960s and 1970s, cross-fertilization of academic work has nurtured the convergence of sub-areas of both. According to Boczkowski and Lievrouw (2008), scholars in communication studies began drawing on theoretical perspectives from technology studies (i.e., interpretative flexibility, social construction of technology, socio-technical systems) while technology studies scholars have increasingly focused on areas of interest within communication studies (i.e., process of technological development processes, media consumption and product, social consequences of technological change) have preoccupied scholars in both STS and communication studies. This cross-fertilization has helped to strengthen both areas in their shared efforts to discern technological meaning in life and society. According to Boczkowski and Lievrouw (2008):

For STS, communication studies has provided an extensive body of social science research and critical inquiry that documents the relationships among mediated content, individual behavior, social structures and processes, and cultural forms, practices, and meanings. For communication studies, STS has provided a sophisticated conceptual language and grounded methods for articulating and studying the distinctive sociotechnical character of media and information technologies themselves as culturally and socially situated artifacts and systems (p. 950).

The crossover of scholarly work between technology studies, information science, and communications is important to Information and Communication Technoethics in key ways. For one, it helped create a substantial interdisciplinary body of research for dealing with new problems and issues nurtured in by advancing ICT's and their consequences. In addition, important areas of crossover between technology studies and communications gave rise to scholarly work on ethical and social implications (which form the base for communication technoethics). Key areas of crossover with an ethical and social orientation emerged in media studies, communication theory, and information theory, and science and technology studies.

Second, core contributions in media studies emerged in the1960s and 1970s as exemplified by the work of McLuhan (and others). Perhaps more than any scholar of our time, McLuhan helped bring together communication and technology considerations by recognizing the influence of technology (media) on human cognition (sensory awareness) and communication content (message). McLuhan's *The Gutenberg Galaxy: The Making of Typographic Man* (1961) posited that, "If a new technology extends one or more of our senses outside us into the social world, then new ratios among all of our senses will occur in that particular culture." Subsequently, in *Understanding Media: The Extensions of Man* (1964), McLuhan's trademark slogan, "the medium is the message" raised public awareness about the important relationship between technology, human cognition, and communication. This work was of seminal importance in the eventual emergence of communication technoethics because it linked together the study of technology and communication in a new way that had not been done before. Building on this work, a number of research programs beginning in the 1970s delved deeper into technology as a medium closely intertwined within communication, culture and society. This included scholarship on media effects (Meyrowitz, 1985), media culture (McLuhan, 1970), media ecology (Real, 1975;; Williams, 1975), and medium theory (Meyrowitz, 1994), and critical media studies (Postman, 1992). For instance, the cultural turn in media studies rooted in media ecology work (and other efforts) views media environments in a broader framework which includes cultural aspects of media. In addition, attention to media ethics, and professional ethics for journalists and broadcasters has helped raise public awareness of the role of media studies in taking a leadership role in shaping the landscape of media to meet public needs and interests. The expansion of core media studies work has helped sustain its vital role in communication research work while expanding its scope to take on practical real world considerations within the public realm. The added social and ethical aspects of media has helped give the field a greater importance and a stronger role to play in the development and management of media operations and the professionals working within it.

Taken together, it appears that technological advances have acted as a driver of progress and regress. On the one hand, technology has facilitating a transformation of organizations and provided evidence of increased scope and speed of communication an information exchange between individuals and organizations on a local and global level. On the other hand, this has also created major social, ethical, and legal problems, as well as a schism between critics and supporters of technology within various communication contexts. Within contemporary society driven by technological advances and organizational change, communications and technology studies play a vital role in gathering strategic information from organizational members (insider perspectives), generating comprehensive knowledge about challenges within changing organizations, identifying technological abuses and crimes that negatively impact organizations and those affected by them. This has, in turn, created a number of social and ethical challenges in dealing with technology and communication. As will become apparent in the next section, the cross-fertilization of communication and technology studies has been particularly important in leading social and ethical inquiry on connections between technology and ethics within various communication contexts (i.e., Internet environments, organizations, human communications, communication theory)

THE ETHICAL TURN IN ICOMMUNICATIONS, TECHNOLOGY STUDIES, AND INFORMATION STUDIES

The Ethical Turn

Information and Communication Technoethics, as it has evolved from the cross-fertilization of information studies, communications and technology studies (and other interdisciplinary scholarship) deals with social and ethical considerations connected to technology within information communication contexts . On the media and theory (or technology) side, Information and Communication Technoethics is rooted in media and technology studies which focused on the complex and changing relationships between humans and technology (i.e., Internet Ethics, Cyberethics, Communication Theory). On the information and communication side, Information and Communication Technoethics is connected to the development of advanced ICT's and how they are used to advance different aims. Given the tremendous impact of the information revolution on society, along with rapid development of new ICT's, it is not surprising that ethical aspects of information and communication began to take on an importance within information science, communication, and technology studies (Tavani, 2007).

Internet Ethics and Cyberethics

In the 1980s and 1990s, advances in computer technologies, early scholarship in Internet Ethics, and creation of codes governing ethical Internet use helped pave the way for present day work in Internet Ethics and Cyberethics. Internet Ethics continued to gather momentum throughout the 1990s while nurturing in a variety of new scholarship under the umbrella of Cyberethics. Work in Cyberethics focused on emerging ethical issues that coincide with the use and misuse of new digital technologies in controversial areas of ethical concern including, privacy rights, Internet censorship, intellectual property issues, online music piracy, and cybercrime (e.g., online identity theft, online fraud, cyberstalking, etc.) These key developments opened up a great deal of additional interest in social research on the Internet that, as will be shown in this chapter, continues to evolve.

Internet Ethics and Cyberethics constitutes an important branch of technoethical inquiry contributing to an amassing body of Internet research concentrating on ethical conflicts and debates. Examples of recent scholarship in this area include, cyberdemocracy (Ribble & Bailey, 2004), censorship issues (Regan, 1996), cyberstalking (Adams, 2002), Internet pornography (Jenkins, 2001), network attacks (Wilson, 2005), information theft, unauthorized transmission of confidential information (Redding, 1996), transmission of offensive jokes, deceptive communication, exchanging online pornography, and cyber identity theft (Roberts, 2008). Typical questions in Internet Ethics and Cyberethics include, "What constitutes Internet abuse and how can this be resolved if at all? How can work organizations deal with personal use of the Internet during work hours to engage in unethical activities such as pornography and hate speech?", "Who should have access and control over the Internet infrastructure?", "What interventions could help safeguard Internet users from unnecessary risks derived from Internet use?", "What are the ethical responsibilities of Internet researchers regarding the use of personal information on the Internet?" and "What are the ethical responsibilities of Internet provides to protect the identity and confidentiality of data on the Internet?

Communication Theory and Technoethical Perspectives

A number of advances in communication theory provided technologists and communication experts with new approaches to deal with ethical approaches to technology mediated communication in a variety of contexts (personal, organizational, societal). Broadly based in rationalism, work in argumentation theory (ala Toulmin), and discourse ethics Habermas' (1992) *Moral Consciousness and Communicative Action* provided a leading framework for logical decision making that could be applied within organizations and other complex situations involving multiple actors and competing interests. In reworking Kantian deontological ethics, Habermas shed new light on the relation between communicative structures and universal moral obligations giving rise to communicative rationality. Habermas' principle of discourse ethics states, "Only those norms can claim to be valid that meet (or could meet) with the approval of all affected in their capacity as participants in a practical discourse," (Habermas, 1992). This body of work is important because it was the first substantive effort to ground ethical inquiry within general communication strategies which could be structured within technology mediated environments (online forums, communities of practice, social networking environments, etc.). As indicated by Visala (2008), general ethical communication strategies also apply to communication challenges within organizational contexts:

Within in the organisational development people's arguments rise from their personal or group interests, which in turn are based on the systemic differentiation of society and technology at a given time. We face a crucial issue: Must we accept separated group interests as inescapable circumstances, or can we reach out for universal human interests? (103).

Rational models for ethical communication and decision-making within organizations are a promising area of inquiry that offers a means of ensuring organizational accountability by offering organizational stakeholders (and affected public). This can be used for negotiating organizational activities including, which products and services to develop, how to implement them, when to avoid situations which put humans, animals, and the environment at risk. Rational planning and decision-making models can be used to provide a strategy and framework for making rationally sound decisions within an organization (Robbins and Judge, 2007). This may be directed to a number of organizational challenges (i.e., identifying an ethical problem, creating ethical evaluation standards, implementing solutions, and monitoring progress). For example, the stakeholder theory developed by R. Freeman is one leading ethical theory within organizational ethics which places morals and values at the centre of organizational management. The general aim in stakeholder theory define organizational stakeholders and the conditions under which potential participants should be treated as stakeholders. One of the main challenges is to define stakeholders in the midst of complex influences from a variety of sources, including, government agencies, political groups, trade unions, affiliated organizations, local communities, employees, prospective clients, and the general public (Phillips & Freeman, 2003). Other work from communication theory from Johannesen's (2008)*Ethics in human communication* delves into ethical responsibility issues from a variety of perspectives (political, human nature, dialogical, situational, religious, utilitarian, and legalistic). The text helps demonstrate how ethical dilemmas arise in almost all areas of communication work including interpersonal communication, small group discussion, organizational communication, codes of ethics, intercultural communication, and multicultural communication. Understanding communication ethics and responsibility is particularly important in areas of crisis. This is partly because of the complexity of crisis situations which allows organizational responses often lead to multiple interpretations, and conflicting views between organizational stakeholders concerning crisis evidence, the organization's mission, and the assignment of responsibility.

Information Ethics

Information ethics emerged in the last decade from work in computer ethics with a focus on ethical issues connected to the development and application of information technologies (Floridi, 1999). Information ethics helped draw attention to ethical aspects of information including, informational privacy issues, information ownership and copyright issues, and information access considerations. Information ethics broadly examines issues related to, among other things, ownership, access, privacy, security, and community (Elrod & Smith, 2005). It also raises attention concerning the important role of information technology in affecting fundamental human rights revolving around copyright protection, intellectual freedom, accountability, and security (More, 2005).

Taken together, work at the intersection of information science, technology studies and communications has helped situate the study of technology and ethics within core areas of media (technology) studies, information science, and communication. Thos, in turn, helped ground Information and Communication Technoethics.

CURRENT WORK IN COMMUNICATION TECHNOETHICS

Current work in communication technoethics concerned with ethical issues and responsibilities arising when dealing with information and communication technology (ICT) in the realm of communication (Luppicini, 2008). It includes contemporary work in communication ethics (Habermas, 1992), media ethics (Patterson &

Wilkins, 2004), journalism ethics (Sanders, 2003), information ethics (Flores, 1999) and other ethical inquiries into communication structures and processes within a variety of technological contexts (e.g., broadcast media, film, radio, print media, Internet, various online environments, etc.). To this end, it provides grounding within technoethics to help guide these sub-field specialty areas of scholarship focused on communication and technology.

One of the main driving forces in current communication technoethics revolves around the ongoing convergence of technology and information communications within contemporary society. The ethical and social issues arising from this marriage are nurturing in potent new interdisciplinary work. There are a number of promising areas of current scholarship which have been especially productive in advancing understanding about relationships between technology (and science), culture, and society. At the centre of this work is the notion of technological convergence and its close connection to individuals and society. This is but one area where Information and Communication Technoethics is helping to build new knowledge about the complexity of technology development processes while raising awareness concerning the social and ethical consequences of technological change.

Information and Communication Technoethics assumes that the culture of technological convergence transforms life and society in important ways. This can be divided into the outward turn and the inward turn in technological convergence. The outward turn in technological convergence concerns the recent recognition of the close connections between technology and culture as the ubiquity and centrality of technology becomes entrenched within different social and cultural contexts. For instance, Sterne's (2003) *The Audible Past: Cultural Origins of Sound Reproduction* is an excellent example of current scholarship which views technology and culture as intertwined processes. In traces out a speculative history of sound recording and culture, Sterne examines the links between human culture and various technologies including telegraph, stethoscope, telephone, radio, and phonograph. The text views sound recording at the turn of the century as "a product of a culture that had learned to can and to embalm, to preserve the bodies of the dead so that they could continue to perform a social function after life" (p. 292). In this sense, the history of sound recording is conceptualized as "the history of the transformation of the human body as object of knowledge and practice" (p. 51). This is but way that the role of technology and culture is taking on new meaning within contemporary society marked by technological convergence in various areas of human activities society affected by the media. Jenkins' (2006) *Convergence Culture: Where Old and New Media Converge* provides an insightful argument concerning the complexity and close relationship between media, information and communication infrastructure, organizational dynamics, and the role of culture and power. He describes this unique development in 21st century as indicative of a culture of convergence:

Convergence represents a paradigm shift – a move from medium-specific content toward content that flows across multiple media channels, toward the increased interdependence of communications systems, toward multiple ways of accessing media content, and toward ever more complex relations between top-down corporate media and bottom-up participatory culture. (243)

The culture of technological convergence is further complicated by its inward turn towards technology and human embodiment. A number of recent works have provided promising work to help inform the inward turn in technological convergence (Turtle & Howard, 2002; Mitchell & Thurtle, 2004; Sutherland, Isaacs, Graham, & McKenna, 2008). Turtle and Howard's edited, *Semiotic Flesh: Information & the Human Body* pulls together insightful essays discussing various aspects of the convergence between information and flesh (body). Mitchell and Thurtle's edited (2004) *Data Made Flesh: Embodying Information*, provides another set of essays exploring the links between information and embodiment. In focusing the relation between technology and the senses, this work attempts to demonstrate how closely related technology is to human experience, "The effects of technology do not occur at the level of opinions or concepts, but alter sense ratios or patterns of perception steadily and without any resistance" (8). This work is expected to evolved as the separation of body from information and communication becomes more blurred within the context of technological growth. Sutherland, Isaacs, Graham, and McKenna's (2008) *Towards Humane Technologies: Biotechnology, New Media, and Ethics* provides an excellent synthesis of key ethical issues at the intersection of biotechnology and communications. This book critically examines the interrelationships of biotechnology, new media and ethics by juxtaposing the work of various scholars and addressing current ethical questions in the public eye.

The outward and inward turn in technological convergence raises questions that are far from being resolved. What is the relationship between the human body and ICT's? How do ICT's influence social interactions of human bodies? How do ethical considerations map onto a closer linking of ICT's to the human condition? What are the ethical and political implications of new biotech products accessible to limited numbers of people? What are the ethical dilemmas created by the public acceptance of prenatal screening for eugenic possibilities and the use of genetic information for illness by insurance companies seeking to exclude certain people? What ethical responsibility is required to ensure that the risks and benefits of biotech developments are publically understood?

Current work in Information and Communication Technoethics has extended assumptions of technological convergence in life and society to organizational contexts. In particular, the use of electronic technologies in organizational settings for communication and monitoring is becoming an important area of research and ethical inquiry. Electronic technology common to organizations includes, but is not limited to, computer monitoring and filtering systems, surveillance cameras, IM and other chat tools, electronic mail and voicemail, Internet, audio and video conferencing tools, personal data assistants, and mobile devices. A small but growing body of research has begun exploring the wider implications of the organizational use of electronic technologies including, surveillance cameras (Whittaker, 1995; Zweig & Webster, 2002), and computer monitoring and filtering systems (Alge, 2001; Urbaczewski & Jessup, 2002). Despite the spread of such countermeasures (and the allocation of limited organizational resources), personal use of electronic technologies within educational and workplace contexts during work hours is not declining, but rather, appears to be gaining ground. Interesting view contributed by Matheson (2007) argues that a "surveillance society risks undermining the ability of its citizens to develop virtue for the same sorts of reasons that overprotective parenting can impair the character development of children." Typical questions in communication technoethics include, "To what extent is democratic decision making facilitated or inhibited in online communications?", "How are ethical conflicts in communication influenced by various technological contexts?", "How should virtual organizations deal with communication conflicts between stakeholders?", and "What procedures should be followed by journalists to protect informant identity and confidentiality within various technological contexts? How should surveillance cameras and other surveillance technologies be used ethically within organizational settings and other public spaces?

TECHNOETHICAL FRAMING

An amassing body of Internet research has focused on a number of ethical issues including, cyberdemocracy, censorship, network attacks, information theft, and deceptive communication. It is apparent that the ethical issues arising from the intertwining of technology and communications in this new communication medium has created new ethical challenges that society is struggling to deal with.

When dealing with any technology mediated communication context, there are interests and needs from a variety of sources to consider: the organization desires to use technology mediated communication to leverage efficient operations and reputation building, society requirements that communication and technology advancement adheres to social norms and governmental regulations, individuals desire for ICT services that offer easy access and privacy without fear of exploitation or abuse. The types of decisions made depend on the weighing of technoethical knowledge derived from key relational dimensions. An examination of technology mediated communication provides useful example of how a relational matrix of technoethical knowledge is applied. There is a growing controversy over the use of technology to enhance communications and information exchange.

From a technoethical standpoint, communication tools should be guided by ethical knowledge about the human condition in relation to the unique communication context. Communication leaders should ensure that individuals are not psychologically harmed (online stalking, flaming, etc) in any way (mental) or treated in ways compromise their dignity and well being (Individual). Moreover, communication regulations require strict adherence to existing protocols along with key ethical aspects of organizational rules (organization). Ethical considerations have an important role to play in all technology mediated communication contexts although they may vary in terms of ethical weight within different societal contexts (societal).

DISCUSSION QUESTIONS

- Is online pornography unethical or should it be treated as an expression of free speech and artistic expression?
- What new threats to intellectual property rights does the Internet create? Are there legal or ethical differences between stealing a CD from a music store and downloading music from the Internet without permission or payment?
- Should researchers disclose their identities when viewing items posted on the Internet in the public domain?
- Do you think that the Internet has changed the world of work and society? How?
- How do you think technology has increased or decreased democratic participation in society over the last 20 years?
- What is the link between ethics and corporate compliance? Should policies be created to force companies and their executives to behave ethically?.
- What ethical policies should guide corporate leaders when engaged in e-business outsourcing?
- What are the ethical concerns in outsourcing company work to third parties at the expense of employment opportunities for regular employees?

DISCUSSION

This chapter traced out the intersection of technology studies and communications. It explored recent developments into how the two fields converge with one another and provide mutual grounding for ethical inquiries into technology and communications. It is shown how the convergence in ICT, combined with intersecting interests between varies areas of STS and communication scholarship helped to draw attention to the broad range of social and ethical concerns at the base of communication technoethics. Additional convergences in bioinformatics and biotechnology have intensified ethical concern and provided new directions for work in communication technoethics. The case is made in this chapter that a connection between these two fields has helped expand the scope of technology and communication scholarship to provide a stronger basis for more cogent analyses of ethical aspects of communication and technology in society. First, converging interests in communications and technology studies allow a closer conceptual linking between technology and society within studies of technology development. This helps to broaden the scope of both fields concerning social and ethical aspects of technical change. It also helps these intersecting fields to transcend the boundaries within their separate research programs to benefit from the expertise and insights of both. Second, a relational system focus between fields provides additional grounding for continued work focusing on the consequences of social and ethical aspects of technical change across multiple contexts. This convergence of two fields provides a high capacity for empirical work in a plethora of communication and technology contexts. Finally, this convergence in communication ad technology helps pulls together key developments of sub-areas of communications and technology dedicated to ethics and technology (i.e., media ethics, cyberethics, journalist ethics).

REFERENCES

Adam, A. (2002). Cyberstalking and Internet pornography: Gender and the gaze. *Ethics and Information Technology, 4*, 133–142. doi:10.1023/A:1019967504762

Alge, B. J. (2001). Effects of computer surveillance on perceptions of privacy and procedural justice. *The Journal of Applied Psychology, 86*(4), 797–804. doi:10.1037/0021-9010.86.4.797

Elrod, E., & Smith, M. (2005). Information Ethics. In C. Mitcham (ed.), *Encyclopedia of Science, Technology, and Ethics*, (Vol. 2: D-K, pp. 1004-1011). Detroit: Macmillan Reference USA.

Floridi, L. (1999). Information Ethics: On the Theoretical Foundations of Computer Ethics. *Ethics and Information Technology, 1*(1), 37–56. doi:10.1023/A:1010018611096

Habermas, J. (1989). *Moral consciousness and communicative action*. Cambridge, MA: MIT Press.

Hakken, D. (2003). *The knowledge landscapes of cyberspace.* New York: Routledge.

Internet Architecture Board. (1989). Retrieved June 4, 2007, from http://tools.ietf.org/html/rfc1087University of Chicago Press

Jenkins, H. (2006). *Convergence culture: Where old and new media converge.* New York: NYU Press.

Jenkins, P. (2001) *Beyond tolerance: Child pornography on the Internet.* New York: New York University Press.

Johannesen, R. (2008). *Ethics in human communication* (6th edition. New York: Waveland Press.

Jones, C. (2006). *Sensorium: Embodied experience, technology.* Boston: MIT Press.

Knorr-Cetina, K. (1999). *Epistemic cultures: How the sciences make knowledge.* Cambridge, MA: Harvard University Press.

Latour, B., & Woolgar, S. (1986). *Laboratory life: The construction of scientific facts.* Princeton, NJ: Princeton University Press.

Luppicini, R. (2008). The emerging field of technoethics. In R. Luppicini & R. Adell (Eds.), *Handbook of research on technoethics* (pp. 1-19). Hershey, PA: Information Science Reference.

Marshall, K. (1999). Has technology introduced new ethical problems? *Journal of Business Ethics, 19*(1), 81–90. doi:10.1023/A:1006154023743

Matheson, D. (2007). Virtue and the surveillance society. *International Journal of Technology . Knowledge and Society, 3*(5), 133–140.

McLuhan, M. (1962). *The Gutenberg galaxy.* Toronto: McGraw Hill.

McLuhan, M. (1964). *Understanding media: The Extensions of Man.* Toronto: McGraw Hill.

Meyrowitz, J. (1985). *No sense of place: The impact of electronic media on social behavior.* New York: Oxford University Press.

Meyrowitz, J. (1994). Medium theory. In D. Crowley & D. Mitchell (eds.), *Communication theory today (pp. 50-77).* Stanford, CA: Stanford University.

Mitchell, R., & Thurtle, P. (2004). *Data made flesh: Embodying information.* NY: Routledge.

Moor, J. H. (2005). Why we need better ethics for emerging technologies. *Ethics and Information Technology*, 7(3), 111–119. doi:10.1007/s10676-006-0008-0

Moore, A. (Ed.). (2005). *Information ethics: Privacy, property, and power.* Washington, DC: University of Washington Press.

Nentwich, M. (2003). *Cyberscience: Research in the age of the Internet.* Vienna: Austrian Academy of Science Press.

Nicholls, A., & Opal, C. (2005). *Fair trade: market-driven ethical consumption.* Thousand Oaks: Sage.

Palm, E., & Hansson, S. O. (2006). The case for ethical technology assessment (eTA). *Technological Forecasting and Social Change*, 73, 543–558. doi:10.1016/j.techfore.2005.06.002

Patterson, P., & Wilkins, L. (2004). Media ethics: Issues and cases (5th edition). New York: McGraw-Hill.

Phillips, R., & Freeman, E. (2003). *Stakeholder theory and organizational ethics.* Berlin: Berrett-Koehler Publishers.

Postman, N. (1992) *Technopoly: The surrender of culture to technology.* New York: Vintage Books.

Real, M. (1975). Cultural studies and mediated reality. *Journal of Popular Culture*, 9(2), 81–85.

Redding, W. C. (1996). Ethics and the study of organizational communication: When will we wake up? In J. A. J. M. S. Pritchard (Ed.), *Responsible communication: Ethical issues in business, industry, and the professions (pp. 17-40).* Cresskill, NJ: Hampton Press.

Regan, S. (1996). Is there free speech on the Internet? Censorship in the global Information infrastructure. In R. Shields (ed.) *Cultures of Internet virtual spaces, real histories, living bodies.* London: Sage.

Ribble, M., & Bailey, G. (2004). Digital citizenship: Focus questions for implementation. *Learning & Leading with Technology*, 32(2), 12–15.

Roberts, L. (2008). Cyber-victimisation in Australia: Extent, impact on individuals and responses. *TILES Briefing Paper, 6.* Retrieved November 28, 2008 from http://www.utas.edu.au/tiles/publications_and_reports/briefing_papers/Briefing%20 Paper%20No%206.pdf

Roberts, L. D. (2008a). Cyber-victimization. In R. Luppicini & R. Adell (Eds.), *Handbook of research on technoethics (pp. 575-593)*. Hershey, PA: Information Science Reference.

Roberts, L. D. (2008b). Cyber identity theft. In R. Luppicini & R. Adell (Eds.), *Handbook of research on technoethics (pp. 542-557)*. Hershey, PA: Information Science Reference.

Roberts, L. D., & Parks, M. R. (1999). The social geography of gender-switching in virtual environments on the internet. *Information Communication and Society, 2*, 521–540. doi:10.1080/136911899359538

Roberts, L. D., & Smith, L. M., & Pollock, C. M. (2008). Ethical issues in conducting online research. In M Khosrow-Pour (Ed.), *Encyclopedia of information science and technology* (2nd Ed., pp. 1443-1449). Hershey, PA: Information Science Reference.

Roberts, L. D., Smith, L. M., & Pollock, C. M. (2005). Conducting ethical research in virtual environments. In M. Khosrow-Pour (Ed.), *Encyclopedia of Information Science and Technology*, (pp. 523-528). Hershey, PA: Idea Group Inc.

Rothman, H., & Scott, M. (2004). *Companies With A Conscience*. Denver, CO: MyersTempleton.

Sanders, K. (2003). *Ethics and journalism*. London: Sage Publications.

Shannon, C., & Weaver, W. (1949). *The mathematical theory of communication*. Urbana, IL: University of Illinois Press.

Sterne, J. (2003). *The audible past: Cultural origins of sound reproduction*. Durham, NC: Duke University Press.

Sutherland, N., Isaacs, P., Graham, P., & McKenna, B. (Eds.). (2008). *Towards humane technologies: biotechnology, new media, and ethics*. Rotterdam: Sense Publishers.

Tavani, H. (2007). *Ethics and technology: Ethical issues in an age of information and communication technology*. Hoboken, NJ: John Wiley and Sons, Inc.

Thilmany, J. (2007, September). Supporting ethical employees. *HRMagazine, 52*(2), 105–110.

Tullberg, S., & Tullberg, J. (1996). On human altruism: The discrepancy between normative and factual conclusions. *Oikos, 75*(2), 327–329. doi:10.2307/3546259

Turtle, P., & Howard, P. (Eds.). (2002). *Semiotic Flesh: information and the human body*. Seatle: University of Washington Press.

Urbaczewski, A., & Jessup, L. M. (2002). Does electronic monitoring of employee Internet usage work? *Communications of the ACM, 45*(1), 80–83. doi:10.1145/502269.502303

Vanderburg, W. H. (2005). *Living in the labyrinth of technology.* Toronto: University of Toronto Press.

Visala, S. (2008). Planning, interests, and argumentation. In R. Luppicini & R. Adell (eds.) (2008). *Handbook of research on technoethics (pp 103-110).* Hershey, PA: Idea Group Publishing.

Visser, W., Matten, D., Pohl, M., & Tolhurst, N. (eds.). (2008). *The A to Z of corporate social responsibility.* New York: Wiley.

Wenger, E. (1998). *Communities of practice: learning, meaning, and identity.* Cambridge, UK: Cambridge University Press.

Whittaker, S. (1995). Rethinking video as a technology for interpersonal communications: Theory and design implications. *International Journal of Human-Computer Studies, 42*(5), 501–529. doi:10.1006/ijhc.1995.1022

Williams, R. (1975). *Television: Technology and cultural form.* New York: Schocken.

Wilson, C. (2005). Computer attack and cyberterrorism: Vulnerabilities and policy issues for Congress.

Zweig, D., & Webster, J. (2002). Where is the line between benign and invasive? An examination of psychological barriers to the acceptance of awareness monitoring systems. *Journal of Organizational Behavior, 23*(5), 605–633. doi:10.1002/job.157

ADDITIONAL READING

Beaulieu, A., & Park, H. (Eds.). (2003). Internet networks: The form and the feel. *Journal of Computer Mediated Communication, 8 (*4), special issue. Retrieved February 1, 2009, from http://jcmc.indiana.edu/vol8/issue4/.

Boczkowski, P. (1999). Mutual shaping of users and technologies in a national virtual campus. *The Journal of Communication, 49,* 86–106. doi:10.1111/j.1460-2466.1999.tb02795.x

D'Abate, C. P. (2005). Working hard or hardly working: A case study of individuals engaging in personal business on the job. *Human Relations, 58,* 1009–1032. doi:10.1177/0018726705058501

Donath, J. A. (1999). Identity and deception in the virtual community. In M. A. Smith & P. Kollock (Eds.), *Communities in cyberspace (pp. 29-59).* London: Routledge.

EthicsWorld. (2009). Retrieved February 1, 2009, from http://www.ethicsworld. org/ethicsandemployees/ethicsandemployees.php

Galloway, R., & Thacker, E. (2007). *The exploit: A theory of networks:* Minnesota: University of Minnesota Press.

Gattiker, U., & Kelley, H. (1999). Morality and computers: Attitudes and differences in moral judgments. *Information Systems Research, 10*(3), 233–254. doi:10.1287/ isre.10.3.233

Griffiths, M. (2003). Internet abuse in the workplace: Issues and concerns for employers and employment counselors. *Journal of Employment Counseling, 40*(2), 87–96.

Gurak, L. J. (2001). *Cyberliteracy: Navigating the Internet with awareness.* New Haven: Yale University Press.

Hayles, N. (1999). *How we became posthuman: virtual bodies in cybernetics, literature, and informatics.* Chicago: University of Chicago Press.

Hine, C. (Ed.). 2006. New infrastructures for knowledge production: Understanding e-science. Hershey: Idea Group.

Lavoie, J. A. A., & Pychyl, T. A. (2001). Cyberslacking and the procrastination superhighway: A web-based survey of online procrastination, attitudes, and emotion. *Social Science Computer Review, 19*(4), 431–444. doi:10.1177/089443930101900403

Lehman, M. (2007). *Embodied music cognition and mediation technology.* Boston: MIT Press.

Lim, V. K. (2002). The IT way of loafing on the job: Cyberloafing, neutralizing and Distraction versus destruction. *Cyberpsychology & Behavior, 9*(6), 730–741.

Mastrangelo, P. M., Everton, W., & Jolton, J. A. (2006). Personal use of work computers: organizational justice. *Journal of Organizational Behavior, 23,* 675–694.

Zielinski, S. (2008). *Deep time of the media: Toward an archaeology of hearing and seeing by technical means.* Boston: MIT Press.

Chapter 7
Biomedical Technoethics

INTRODUCTION

It is now argued that people do own and control their bodies (and tissues) but have no right to sell them, for they cannot exist without them and their rights as human beings and as consumers cannot trump the right of society to attempt to stave off the process of co modification of human life. –Fait, 2008, p. 209).

The ethical use of new technologies is important for the advancement of modern health and medicine in many areas including, pharmaceuticals and healthcare, reproductive technologies and genetic research, sports, and nutrition. This is partly because revolutionary technologies have provided new opportunities that raise ethi-

DOI: 10.4018/978-1-60566-952-6.ch007

cal issues and public debate. In this context, technology is viewed not as a solution, but as a driver of societal development fostering change and generating new ethical considerations to address. The goal of technoethical inquiry within the context of modern health and medicine is to ensure that technological changes are desired and accepted by those affected.

Technoethics assumes that advancing work in biotechnology do not exist in a vacuum and are not socially, politically and ethically neutral. Rather, there are complex interrelations to consider between technological innovation, individual interests and needs, societal norms and values, organizational structures, and historical events. Some areas of biotech research are at low risk for creating a negative impact and require little attention outside the domain of biotechnology. Other areas of biotech research at high risk for creating a negative greater impact and may require the establishment of ethics committees, compliance with rigorous ethical standards, and government intervention in the private sphere. This is the particularly true when dealing with areas of biotech research that affect human populations.

As will be discussed in this chapter, the relationship between ethics and technology within the context of modern healthcare and medicine raises a number of concerns in key areas affected by recent technology advances. New life-preserving technologies, advances in stem cell research, and controversial cloning technologies are redefining current conceptions of life, death, and patient care. Advances in nutrition and pharmacology offer new opportunities to enhance performance, extend life, and alter mental and physical states. Ethical issues arising from such advances have created a need for a technoethical framework in health and medicine to deal with a variety of exciting new technology related innovations which are redefining our biological and physical being.

This chapter reviews key technoethical speciality areas guiding the ethical use of technology within the context of modern healthcare and medicine including biotech ethics, and teleheath. By examining key developments, public concerns, and the role of ethics committees, this chapter provides information on key areas of concern and possible applications. By drawing on a technoethical framework, this chapter makes suggestions to help protect the public while guiding new developments in biotechnology that ensure mutual benefit-sharing.

BACKGROUND

The advances of modern medical technology transformed medical practice in the 20th century while creating a need for. First, new instrumentation allowed observation and measurement beyond the capacity of physicians. For instance, in 1913, the electrocardiograph began to be used for a diagnostic tool for exploring heart functioning. In 1935, a heart-lung machine was used to monitor heart and lung functions of a cat during surgery. This was adapted for use on humans in 1953 during open heart surgery. In 1952, the first cardiac pacemaker was created for controlling irregular heartbeat. The first artificial heat was implanted in a dog in 1957 and by 1967, the first human heart transplant was performed. The outcome of sophisticated new instrumentation, not only created a strong relationship between humans and technology in the medical profession that placed technology at the forefront of medical practice, but it also raised ethical considerations to deal with.

Second, genetic research redefined the boundaries of life and death as well as challenging current notions of human capacity and identity. Contemporary interest in genetic engineering can be traced to the 20th century movement in Eugenics. The term "eugenics" derived from the Greek meaning of "wellborn" was first coined by Sir Galton in 1883. In 1912, the first (of three) International Congress of Eugenics defined eugenics as "the self direction of human evolution." The main thrust behind this movement was to promote the advancement of human hereditary traits through intervention. This included both negative eugenics (efforts to eliminate inheritable diseases and malformations via various techniques including birth control, strategic abortion, sterilization, immigration control, and in some cases, euthanasia) and positive eugenics (promoting the proliferation of desirable mental and physical traits via monetary incentives for selected parents, and human mate selection). This early work set the stage for modern day genetic research. Contemporary work in genetic research began in 1953 with James Watson and Francis Crick who first discovered the structure of the DNA molecule. This work continued to evolve through the Humane Genome Project and cloning research. In1996, the biggest breakthrough in cloning research in the 21st century occurred when a sheep (Dolly) became the first mammal cloned from an adult cell. The evolution of eugenics and genetic research initiative in contemporary science and technology includes work in reprogenetics, germ line engineering, and cloning, along with mounting public debate concerning the co modification of healthcare, genetic responsibility, genetic discrimination, and genetic enhancement.

Major breakthroughs in 20th century medicine and genetic research were paralleled by an equally important growth in technoethical inquiry that would become known as biotech ethics. One main question that arise is what is the limit of genetic and personality enhancement that society will tolerate? Biotech Ethics grew out of the cross fertilization of Bioethics and Medical Ethics in the mid-1960s and early 1970s. This occurred at a time when the technological advances in medical instrumentation (i.e., dialysis machines, artificial hearts) was already beginning to transform medicine and how it was managed. This technoethical development occurred partly in response to the limited availability of new technologically enhanced medical treatments and high cost of treatment which created accessibility issues. Besides, these practical problems and limitations, a number of new ethical issues arose revolving around controversial areas affected by debates surrounding genetic research and new technologies including, eugenics, abortion, animal rights, euthanasia, *in vitro* fertilization, and reproductive rights emerging in the 1970s and 1980s.

Throughout the 1980s and 1990s, ethical issues continued to arise concerning biotechnology in agricultural applications and food science, thus adding to the momentum behind biotech ethics. Ethical issues surrounding biotechnologies also became increasing important in healthcare, medical research, and various applications in biotechnoloy. Jonas's *On Technology, Medicine and Ethics* (1985) was an early work in technoethical inquiry focusing on new ethical problems in Medicine created by the use of advanced technology. In genetic research, mixed public response to the Human Genome Project led by the U.S. Department of Energy in 1986 raised a number of ethical questions and public concern about the use of technology to map out the sequence of the human genome. Concurrent research altering matter at the level of atoms and molecules, called nanotechnology, was also raising public concern. Drexler's (1986)*Engines of Creation: The Coming Era of Nanotechnology* examined the contributions and risks in nano-scale research and development. Nano-technology research and its applications gradually became a focus of technoethical inquiry (later called nanoethics) due to rising public concern about potential health issues, safety concerns, and environment risk. By the early 1990s, technoethical inquiry broadened in scope in response to decades of progress in various areas of biological technology (i.e., genetically modified food, gene therapy, and stem cell research).Subsequently, in the early 2000s, neuroethics began to emerge as a response to controversial areas of neuroscience which spurred public debate (i.e., pharmacology, neural implantations, neuro enhancement, brain engineering). The rapid progress in research and development in controversial areas of biotechnology has contributed an amassing body of work begin dealing with the ethical and legal implications.

BIOTECHNOLOGY AND THE ETHICAL TURN

The Ethical Turn

Biotech Ethics (and related programs of ethical inquiry) has distinguished itself as an important branch of Technoethics. This is largely due to the continued evolution of biotechnology and specialty areas of research in medicine, biotechnology, healthcare, nanotechnology, and neuroscience. These technology led areas have contributed new work in technoethical inquiry under the general heading of biotech ethics and related specialist areas of inquiry. This section covers recent developments in Biotech

Ethics I key areas, including, genetic research, biotech research ethics, nanoethics, neuroethics, ethics in pharmacology and sport, and tale-health ethics.

The Genetic Turn and the Co Modified Life

There are a number of continuities and discontinuities in current genetic research that raise serious concerns. One major concern about genetic engineering revolves around when to draw the line between gene therapy and species enhancement. In the event that pre-implantation genetic diagnosis (PGD) tests become widespread, embryos with major genetic disorders can be replaced with healthy ones in the womb. The question is what constitutes a major genetic disorder? A second concern is about the ability of PGD to deal with the complexity of genetic disorders. Since many gene disorders are complex and involve a variety of genetic and non-genetic variables, PGD tests are limited in their ability to detect problems. Moreover, there is risk of unanticipated consequences that are damaging and not reversible. A third problem arises when considering the long-term societal influence of unregulated genetic engineering and how it affects individual rights and values. For instance, how will genetically enhanced and non-enhanced humans be treated in society in terms of opportunities and rights?

One new area of both legal and ethical concern revolves around biotech research and the co modification of life. In a time of advancing biotech research, new opportunities for creating and exchanging genetic material have given rise to a new industry while provoking global debate concerning the control and regulation of biogenetic goods and services. What are the rules governing human embryo creation in research and industry? Who are the owners of cell-lines and who should share in the profits from their commercialization? What are the ethical implications of assigning an economic value to bodies? To what extent are patent claims for intellectual property of DNA morally and legally justifiable? To what extent should market forces be permitted to create and sustain markets which place a value on human genetic material? Do what extent does such an industry risk degrading societal conceptions of personhood human life?

In 1997, the European Human Rights and Biomedicine Convention took the position again the creation of human embryos for economic gain. Article 21 stated that "organs and tissues proper, including blood, should not be bought or sold or give rise to financial gain for the person from who they have been removed or for a third party, whether an individual or a corporate entity such as, for example, a hospital." The exception to this is body material that is unneeded and discarded (i.e., hair or placentas). This does not apply to donated transplantation of body material within a formalized system. What is interesting about this position is that it set a precedent that placed limits on individuals' rights to control their own bodies.

Biotech Research Ethics

At the crossroad of ethical concerns and debate about biotech research and engineering lies the research community struggling to continue work in this area. In line with guidelines posited by the Human Rights and Biomedicine Convention, a number of other guidelines have been posited by various organizations to help guide researchers in this controversial arena of humans and advanced technology: World Medical Association (Helsinki Declaration 2000), Council for International Organizations of Medical Sciences (CIOMS 2002), Nuffield Council on Bioethics (2002) and the UNESCO Declaration on Bioethics and Human Rights (2005).

At the core of this concerted effort to tighten research policies in this area lies the ethical concern over possible exploitation of populations in developing nations who are at the highest risk of exploitation. For instance, the Global Forum for Health Research (2004) reported that less than 10% of monetary investment in health research in many developing countries is directed to the advancement of health research in areas that affect the test population. This raises serious concerns because there are different health problems and needs around the world to be addressed that are not being addressed. Compounding this problem is the fact that many of the health problems in developing countries (e.g., infectious diseases, famine, hygiene, etc) can be treated with inexpensive types of medicinal intervention outside the scope of expensive types of, medical research currently conducted in developing countries by industrialized nations (I.e.,. biotech research, reprogenics, etc.). Given that much of the global research activities in developed countries do not contribute to the relief of suffering in these countries, what are the ethical implications of current and future researchers working in this area?

Taking a technoethical perspective entails finding a balance between the needs of developing host countries and the countries conducting research within these host countries. This requires mutual understanding of research risks and benefits from all research stakeholders and those affected by the research. This is echoed in the 2005 UNESCO Declaration on Bioethics and Human Rights in arguing for the implementation of a principle of "benefit sharing" so that less developed countries do not lose out on the benefits of biomedical research (UNESCO Declaration 2005). A technoethical framework focuses on the structural and procedural conditions for establishing a balanced set of ethical relations within the technological context of biotech research. First, a set of procedures to bring stakeholders and other affected groups together is needed to help guide perspective sharing and participatory decision making with the aim of finding a relational balance which satisfies ethical research standards held by all stakeholders and affected groups. Second, some form of institutionalized or standardized structure is needed that enforces these procedures to bring stakeholders and other affected groups together to negotiate ethical research standards that takes all stakeholders and affected groups into consideration. One option is the establishment of strategic research ethics committees in areas of biotech research which are controversial (I.e.. where there is a history abusing research participants, power imbalances between stakeholder groups and those affected by the research, high risk of exploitation, high levels of inequality in benefit sharing). Bagheri (2008) provided the following list of technoethical conditions to help research ethics committees deal with externally-sponsored research in developing countries: (1) research should be relevant to the population in which it is conducted, (2) the relevance of research to the health need of the society is another important issue which has to be determined by the research ethics committee, and (3) the research ethics committees should determine the priority for collaboration in the projects that align with national health policies of the respective countries. Although, technoethics is applicable to all areas of research involving advanced technologies, the role of research ethics committees with a technoethical focus is particularly important when dealing with developing countries, since many do not have the resources or capacity to assess national health research priorities without some assistance. In such situations, the research ethics committee may play a pivotal role in safeguarding ethical research, ensuring that externally-sponsored research conducted in developing countries supports national research priorities in developing countries, and bringing all stakeholders and affected groups together on a variety of key ethical questions that may arise. Should a research ethics committee give more weight to avoiding harm to the subjects or more benefits to the larger population in their risk-benefit assessment? How will the researchers protect research participants and minimize the risks within the specific research context?

The Rise of Neuroethics and Nanoethics

Neuroethics is an outgrowth of biotech ethics and exams the ethical aspects of neuroscientific advances in medicine and biological research. Part of the importance of this new area derives from the fact that it focuses on human consciousness, selfhood, and other mental and behavioural functions of neurological processes that are core to the human condition. The centrality of neurological processes to human life raises unique ethical questions concerning the use and misuse of neuroscientific advances to alter the human mind and body in powerful ways. The appearance of scholarly work in neuroethics coincides with the rise of neuroscience, which preceded any formal recognition of the term label "neuroethics." Steven Marcus' (2002) *Neuroethics: Mapping the Field* was the first major reference work for neuroethics under this label. It pulled together a diverse collection of readings exploring early debates in neuroethics. Gazzaniga's (2005) *The Ethical Brain* helped to formalize the study of neuroethics as "the examination of how we want to deal with the social issues of disease, normality, mortality, lifestyle, and the philosophy of living informed by our understanding of underlying brain mechanisms". The next major development to ground technoethical inquiry derives from the recent establishment of the Neuroethics Society in 2006. This society is made up of an international group scholars dedicated to "promote the development and responsible application of neuroscience through better understanding of its capabilities and its consequences." (Neuroethics Society, 2008). Finally, in 2008, a journal called Neuroethics was created to bring together interdisciplinary research in neuroethics and related areas of the sciences of the mind. Key areas covered in this first year of publication include, ethical aspects of cognitive enhancement, ethics of prescription drug use, neuroscientific discovery and ethical practice, neuroscience and free will, and neuroscience and the question of selfhood (Neuroethics, 2008). These recent developments help secure neuroethics as a promising area of technoethical inquiry. Building on Luppicini (2008), some key questions in this area include, "Who should have ownership and control of harvested DNA, human tissue and other genetic material?", "How should information from genetic tests be protected to avoid infringing on human rights?", "How do we deal with the fact that developing countries are deprived the same access to applications of biotechnology (biotech divide) available in other countries? What is the difference between treating a human neurological disease and simply enhancing the human brain? Another such question might be: Is it fair for the wealthy to have access to neurotechnology, while the poor do not? Neuroethical problems could complement or compound ethical issues raised by genomics, genetics, and human genetic engineering (see Gattaca argument). How safe are the drugs used to enhance normal brain function?

Nanoethics represents another promising new area of technoethical inquiry concerned with ethical and social issues associated with developments in nanotechnology. It has advanced specialty research areas in Computer Science, Engineering, Biology, and Chemistry. Hunt and Mehta (2006) reviewed major developments in nanotechnology research giving rise to public concern over risks and ethical issues associated with this type of technological innovation. Also, in 2007, a new journal called, *NanoEthics: Ethics for Technologies that Converge at the Nanoscale,* was created to address ethical issues related to nanotechnology. The inaugural issue presented new work on a variety of topics including country specific scientist perspectives on nanosciences and nanotechnologies, nanoadministration, nanoregulation, ethics of risk analysis, and nano-enabled diagnostics (NanoEthics, 2007). This increased attention to nanotechnology in technoethical inquiry promises to advance knowledge about ethical and societal concerns arising from nanotechnology research and development. Building on Luppicini (2008), some key questions of interest in this area include, "What are the potential health and safety risks with nanotechnology applications and who is responsible?"," What are the rights of individuals in affecting nanotechnology applications?", and What responsibilities do we have to protect society from the risks of nanotechnology advancement?"

Ethics in Pharmacology, Sports and Nutrition

The rapid growth of pharmacology (the study of the actions of drugs and how they interact with biological organisms) and the pharmaceutical industry in the 20th century has played an important role in transforming medical research and practice at a global level. As with most (if not all) technological innovations in human activity, the influence of the pharmaceutical industry extends far beyond the scope of medicine. There are also legal and safety implications connected to consider. At a political level, government regulation has been created to control the development, sale, and administration of medication. For instance, the approved of prescription drugs in the U.S. is determined by the federal Prescription Drug Marketing Act of 1987 enforced by the Food and Drug Administration (FDA). The FDA is the body that regulates pharmaceuticals by ensuring that approved drugs are demonstrated to be effective against the disease for which it is seeking approval and that they satisfy standards assessed through proper animal and human testing (Rang, Dale, Ritter, & Flower, 2007). Despite the necessity of public safety regulations, this does nothing to address public concerns with social and ethical implications connected to how the pharmaceutical industry operates and exerts influence in our lives. Ethical controversies have erupted around the world concerning the overuse and misuse of certain drugs to treat anxiety, depression, hyperactivity, sexual dysfunction, not to mention, the application of pharmaceutical technology to develop potent sport enhancing drugs.

The controversial use of pharmaceutical intervention in amateur sports through performance enhancing drugs and gene doping is perhaps the most debated area of pharmaceutical concern in society since the Canadian sprinter Ben Johnson was disqualified for illegal steroid use when breaking the world record in the 100metre sprint. This raised the ethical challenge for the world about how to deal with the ethical and unethical role of technological development in sport and nutrition? Miah's (2004) Genetically Modified Athletes: Biomedical Ethics, Gene Doping & Sport' provides a recent sketch of some major areas of this debate raging on in sport technoethics. This debate is important because it extends controversies about the ethical application of genetic research and pharmaceuticals to an area of life and society of global interest--sports.

World Anti-Doping Agency was created in 1999 to act as the main organization responsible for regulating anti-doping policy worldwide. The World Anti-Doping Agency created a code intended to define and control unacceptable pharmaceutical and medical intervention in Olympic sporting events that would compromise values

Figure 1. World anti-doping network for sports

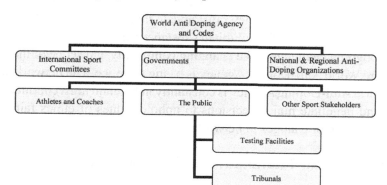

of ethics fair play and honesty, such as gene doping and the use of unauthorized sport enhancing. Gene doping refers to the 'non-therapeutic use of cells, genes, genetic elements, or of the modulation of gene expression, having the capacity to improve athletic performance. (World Anti-Doping Agency, 2003).

Breaking the World Anti-Doping Network for Sports down into core structures and functions reveals its relational complexity involving: (1) The World Anti Doping Code, (2) international sport committees, (3) governments, (4) national and regional anti-doping organizations, (5) testing services, (6) tribunal.

As depicted in Figure 1, The World Anti Doping Agency and Code is the official international organization responsible for preventing the unethical use of technology (doping in sports) in Olympic sport competitions through cooperation within a network of stakeholders and others affected by doping in sport. The international sport committees, such as the International Olympic Committee, International Paralympics Committee, and International Sports Federation, are network partners responsible for testing athletes and enforcing compliance with the codes. Governments are responsible for allowing national testing programs for reducing doping in sport and prohibiting the distribution of products and technologies that are not in compliance with the code. National and regional anti-doping organizations are network partners responsible for testing athletes and enforcing compliance with the codes on a regular basis (beyond the context of Olympic competition). Testing services are network partners established around the world for examining doping tests conducted under the code. Tribunals are investigations carried out within an independent agency to settle sport related legal and ethical disputes according to the codes and procedures held within the world of sports.

Despite the apparent success of efforts from the World Anti-Doping Agency, the complexity of the situation (and the network of individuals affected) and the constant evolution of technology creates an ongoing techneothical struggle with no end in sight. First, the list of unauthorized sport enhancing drugs is continually evolving and the World Anti-Doping Agency must update their research efforts and publish a new list each year to identify prohibited substances and prohibited methods. Adding to this ongoing challenge is the changing complexity and scope of the world of sport that the World Anti-Doping Agency is trying to police. The network affiliated with the World Anti-Doping Agency is constantly in flux and must continually share feedback and negotiate major decisions to uphold values of ethics fair play and honesty in the face of changes in network membership and values.

The Turn Towards Telehealth and Telemedicine

The Institute of Medicine (IOM) defines telehealth (telemedicine) as, "the use of telecommunications and information technologies to share and to maintain patient health information and to provide clinical care and health education to patients and professionals when distance separates the participants. (Field, 1996, p. 27). Breaking the definition of telehealth down into core structures and functions reveals its relational complexity involving: (1) ICT utilization, (2) geographic distance, (3) health or medical interventions, (4) clients (patients and healthcare professions), and (5) service providers.

The complexity of telehealth is made apparent when considering areas of telehealth that have received increasing public attention and debate such as implantable biosensor technology. Biosensors are ICT devices implanted or attached to patients to monitor patient health and link them to telehealth service providers at a distance. According to Bauer (2008):

Implantable biosensors are being used to monitor patients. For example, implantable cardiac biosensors that use wireless technology are being linked to sophisticated Internet-based monitoring networks that allow patients to transmit device and physiologic data to their providers without leaving their homes. Providers can remotely monitor the condition of their patients by logging into a secure website. Patients may also have access to the same website where they can obtain health-related information and personalized device data (170).

On the one hand, implantable biosensor technology has the potential to transform healthcare for many patients by offering patients more continuous healthcare at a distance with a higher degree of patient flexibility in how and where they receive key services. If developed properly, implantable biosensors may make healthcare more proactive rather than reactive. Bauer (2008) believes that the proper use of biosensor technology may advance medicine by resituating [at least some] healthcare delivery from institutional settings to the home, thus giving patients more autonomy and personal control in managing their healthcare. On the other hand, implantable biosensor technology opens up a variety of technoethical questions that have not been addressed. For instance, what are the ethical implications of biosensor technology? Will medically underserved populations gain equitable access to biosensor technology? What are the ethical implications of biosensor technology for provider-patient relationships? What are the ethical and legal implications of biosensor technology for medical privacy and patient confidentiality?

TECHNOETHICAL FRAMING

Based on the discussion of biotechnology and its many ethical aspects, it is clear that technology has changed the world and the role of decision makers responsible for guiding developments that have broad ethical considerations. When working on any technology initiative, there are interests and needs from a variety of sources to consider: the organization desires an efficient operation; society requires that biotechnology advancement adheres to social norms and governmental regulations, individuals desire adequate medical attention and access to life preserving and life enhancing services that are reasonably priced. The types of decisions made depend on the weighing of technoethical knowledge derived from key relational dimensions.

From a technoethical standpoint, biotech research should be guided by ethical knowledge about the human condition in relation to the unique biotech research context. This research should ensure that research participants are not physically harmed in any way (biological) or treated in ways compromise their dignity and well being (Individual). Moreover, biotech research requires strict adherence to regular research protocols along with key ethical aspects of biotech research (organization). Ethical aspects have an important role to play in all biotech research although they may vary in terms of ethical weight within different societal contexts (societal). Drawing on a review of available documentation of biotech ethical research guidelines, Table 1 provides a breakdown of key ethical aspects of biotech research with corresponding guiding questions.

A second example of technoethical framing can be illustrated in the area of Telehealth where a number of key ethical and legal questions can be discerned to help frame a proper technoethical inquiry. Based on the abovementioned work in telehealth, it can be seen that many aspects of telehealth are complex and embedded in a variety of human-technology relations. Based on the chapter review of current biotech research and other relevant literature, Table 2 identifies key ethical aspects of telehealth services for guiding a technological inquiry.

DISCUSSION QUESTIONS

- What are the ethical considerations in telehealth? Do you think medically un-derserved populations gain greater access to healthcare services? What sorts of tradeoffs (if any), between access and quality might occur? What are the ethi-cal implications of telehealth for provider-patient relationships? How might this influence medical information privacy and patient confidentiality?

Table 1. Key ethical aspects of biotech research

Aspects	Guiding Questions
Scientific validity	Is the biotech research using human research participants scientifically sound?
Ethical review	Does the biotech research demonstrate to the ethics review committee that all necessary steps to safeguard the welfare and rights of the research participants and the community is followed?
Informed consent	Have research participants been fully informed about the research project and of the risks involved?
Non-exploitation	Is participant involvement in biomedical research free from discrimination based on education, social status, or wealth?
Benefit & precaution	Does the biotech research ensure that it offers potential benefits and that re-search participants are not placed at more than minimum risk
Privacy	Does the biotech research ensure research confidentiality and guarantee non disclosure of identity or personal data without written consent of research participants?
Compliance and public interest	Does the biotech research comply with accepted research guidelines, local regulations, and established public interest?
Research participant treat-ment and compensation	Does the biotech research ensure offer free medical treatment and compensa-tion for accidental injury caused by the research?

Table 2. Key ethical aspects of telehealth services

Aspects	Guiding Questions
Accessibility	Are all clients and potential clients able to receive telehealth services when needed without substantial burden or delay?
Quality	Are the available telehealth services of high quality and delivered effectively to clients in a compassionate manner?
Fairness	Are the delivery telehealth services free from discrimination based on educa-tion, race, social status, or wealth?
Benefit & precaution	Do the telehealth services offer tangible benefits that do not place clients at more than minimum risk
Privacy	Do the telehealth services ensure confidentiality and guarantee non disclosure of identity or personal data without written consent of clients?
Compliance and public interest	Does the telehealth service network comply with accepted healthcare guide-lines, local regulations, and established public needs and interests?

- Do you believe that telehealth services should replace or supplement in-person services in your country? If so under what conditions? Should it be a public service, private service, or a combination of both? J
- What do you believe are the ethical duties of individuals with regard to their accessible genetic information? (e.g., is there a duty to warn family of potential genetic risks?
- Medical research is often conducted by developed nations in developing countries. Should the products of clinical work (e.g., new therapeutic techniques, drugs, etc,) be made available (to the local community in which the trial was conducted or not?
- There are the major concerns about human genetic research from a social and ethical standpoint? Are these concerns justified or not? What level of genetic and personality enhancement do you believe should be tolerated?
- Future advances in germ-line genetic engineering may result in individuals being born with genetically enhanced predispositions that may provide unfair advantages in competition. There is currently no way of dealing witch such cases under the World Anti-Doping Code. How do you think sport authority should deal with the situation if this does occur?

DISCUSSION

This chapter provided a look at key developments connected to the ethical use of technology within the context of modern healthcare and medicine. By examining key developments, public concerns, and the role of ethics committees, this chapter provided information on areas of concern. By drawing on a technoethical framework, this chapter makes suggestions to help protect the public while guiding new developments in biotechnology that ensure mutual benefit-sharing. A technoethical framing is provided to highlight important ethical aspects to consider in the areas of biotech research and telehealth services.

REFERENCES

Allhoff, F., & Lin, P. (Eds.). (2008). *Nanotechnology and society: Current and emerging ethical issues.* New York: Springer.

Allhoff, F., Lin, P., Moor, J., & Weckert, J. (Eds.). (2007) *Nanoethics: The ethical and social implications of nanotechnology.* New York: John Wiley and Sons.

Anderson, A., Allan, S., Petersen, A., & Wilkinson, C. (2005). The framing of nanotechnologies in the british newspaper press. *Science Communication, 27*(2), 200–220. doi:10.1177/1075547005281472

Anderson, A., Allan, S., Petersen, A., & Wilkinson, C. (2008) Nanoethics: News media and the shaping of public agendas. In R. Luppicini & R. Adell (eds.), *Handbook of research on technoethics* (pp. 135-150). Hershey, PA: Idea Group.

Anderson, A., Petersen, A., Wilkinson, C., & Allan, S. (in press). *Nanotechnology, risk and communication*. Houndmills, UK . *Palgrave Macmillan*.

Bagheri, A. (2008). Ethics review on externally-sponsored research in developing countries. In R. Luppicini & R. Adell (eds.) *Handbook of research on technoethics* (pp112-124). Hershey, PA: Idea Group.

Bauer, K. (2008). Healthcare ethics in the information age. In R. Luppicini and R. Adell (eds.), *Handbook of research on technoethics,* (pp170-185). Hershey, PA: Idea Group.

Colvin, V. L. (2003). The potential environmental impact of engineered nanoparticles. *Nature Biotechnology, 21*(10), 1166–1170. doi:10.1038/nbt875

Council for International Organizations of Medical Sciences (CIOMS) in collaboration with the World Health Organization. (2002). *International ethical guidelines for biomedical research involving human subjects*. Retrieved December 1, 2008, from http://www.cioms.ch/frame_guidelines_nov_2002.htm

Declaration, U. N. E. S. C. O. (2005). *Universal declaration on bioethics and human rights*. Retrieved December 1, 2008, from http://portal.unesco.org/en

Drexler, E. K. (1986). *Engines of creation: the coming era of nanotechnology*. New York: Anchor Books.

ETC. (2003). *The big down: atomtech – technologies converging at the nano-scale*. Winnipeg: Canada.

Fait, S. (2008). Ethical aspects of genetic engineering and biotechnology. In R. Luppicini & R. Adell (eds.), *Handbook of research on technoethics,* (pp145-161). Hershey, PA: Idea Group.

Field, M. (Ed.). (1996). *Telemedicine: A guide to assessing telecommunications in health care*. Washington, DC: National Academy Press.

Friends of the Earth Australia. (2008) *Out of the laboratory and on to our plates: Nanotechnology in food and agriculture*. Friends of the Earth Australia, Europe and USA. Retrieved April 22 2008 from http://nano.foe.org.au/filestore2/download/228/Nanotechnology%20in%20food%20and%20agriculture%20-%20text%20only%20version.pdf

Gazzaniga, M. (2005). *The ethical brain.* Washington, DC: The Dana Press.

Global Forum for Health Research. (2004). *The 10/90 report on health research*, (fourth report). Retrieved December 1, 2008, from http://www.globalforumhealth.org/Site/000_Home.php

Gorss, J., & Lewenstein, B. (2005). *The salience of small: nanotechnology coverage in the american press, 1986-2004*. Paper presented at International Communication Association Conference, May 27th.

Handy, R., & Shaw, B. (2007). Toxic effects of nanoparticles and nanomaterials: Implications for public health, risk assessment and the public perception of nanotechnology. *Health Risk & Society, 9*(2), 125–144. doi:10.1080/13698570701306807

Handy, R., von der Kammer, K., Lead, J. R., Hasselov, M., Owen, R., & Crane, C. (2008). 'The ecotoxicology and chemistry of manufactured nanoparticles. *Ecotoxicology (London, England), 17*, 287–314. doi:10.1007/s10646-008-0199-8

Helsinki Declaration. (2000). Retrieved December 1, 2008, from http://www.wma.net/e/policy/b3.htm

Hoet, P. H. M., Bruske-Hohlfield, I., & Salata, O. (2004). Nanoparticles – known and unknown health risks. *Journal of Nanobiotechnology, 2*(12). Retrieved from http://www.jnanobiotechnology.com/content/2/1/12

Hunt, G., & Mehta, M. (Eds.). (2006). *Nanotechnology: Risk, ethics and law*. London: Earthscan.

Jonas, H. (1985). *On technology, medicine and ethics*. Chicago: Chicago University Press.

Lee, R., & Jose, P. D. (2008). Self interest, self restraint and corporate responsibility for nanotechnologies: Emerging dilemmas for modern managers. *Technology Analysis and Strategic Management, 20*(1), 113–125. doi:10.1080/09537320701726775

Lewenstein, B. V. (2006). What counts as a social and ethical issue in nanotechnology? In J. Schummer & D. Baird, (eds.) *Nanotechnology challenges: Implications for philosophy, ethics and society* (pp. 201-16). London: World Scientific.

Macoubrie, J. (2006). Nanotechnology: Public concerns, reasoning and trust in government. *Public Understanding of Science (Bristol, England)*, *15*(2), 221–241. doi:10.1177/0963662506056993

Marcus, S. (Ed.). (2002). *Neuroethics: Mapping the field.* San Francisco: Dana.

Miah, A. (2004). *Genetically modified athletes: Biomedical ethics, gene doping and sport.* New York: Routledge.

Michelson, E. S., & Rejeski, D. (2006). Falling through the cracks? Public perception, risk, and the oversight of emerging nanotechnologies. *IEEE Report.*

Mnyusiwalla, A., Daar, A. S., & Singer, P. A. (2003). Mind the gap: Science and ethics in nanotechnology. *Nanotechnology*, *14*, 9–13. doi:10.1088/0957-4484/14/3/201

Nanoethics. (2007). *Journal home.* Retreived October 1, 2008, from http://www.springerlink.com/content/120571/?Content+Status=Accepted

Neuroethics. (2008). *Journal home.* Retreived October 1, 2008, from http://www.springer.com/philosophy/ethics/journal/12152

Neuroethics Society. (2008). *Home.* Retreived October 1, 2008, from http://web.memberclicks.com/mc/page.do?sitePageId=33808&orgId=ns

Nuffield Council on Bioethics. (2002). *The ethics of research related to healthcare in developing countries.* London: Nuffield Council on Bioethics.

Petersen, A., Allan, S., Anderson, A., & Wilkinson, C. (in press). Opening the black box: Scientists' views on the role of the mass media in the nanotechnology debate. *Public Understanding of Science (Bristol, England).*

Petersen, A., & Anderson, A. (2007). A question of balance or blind faith? Scientists' and policymakers' representations of the benefits and risks of nanotechnologies. *NanoEthics*, *1*, 243–256. doi:10.1007/s11569-007-0021-8

Preston, C. (2006). The promise and threat of nanotechnology: Can environmental ethics guide us? In J. Schummer & D. Baird (eds.), *Nanotechnology challenges: Implications for philosophy, ethics and society,* (pp. 217-48). London: World Scientific.

Royal Society and Royal Academy of Engineering. (2004). *Nanoscience and nanotechnologies: opportunities and uncertainties report.* London: The Royal Society.

Scheufele, D. A., Corley, E. A., Dunwoody, S., Shih, T., Hillback, E., & Guston, D. (2007). Scientists worry about some risks more than the public. *Nature Nanotechnology*, *2*(12), 732–734. doi:10.1038/nnano.2007.392

Schummer, J., & Baird, D. (Eds.). (2006). *Nanotechnology challenges: Implications for philosophy, ethics and society*. London: World Scientific.

Wilsdon, J., & Willis, R. (2004). *See-through science: Why public engagement needs to move upstream*. London: Demos.

Wood, S., Geldart, A., & Jones, R. (2008). Crystallizing the nanotechnology debate. *Technology Analysis and Strategic Management, 20*(1), 13–27. doi:10.1080/09537320701726320

Wood, S., Jones, R., & Geldart, A. (2007). *Nanotechnology, from the science to the social: The social, ethical and economic aspects of the debate*. Swindon, UK: Economic and Social Research Council.

World-Anti Doping Code. (2003). Retrieved October 1, 2008 from, http://www.wada-ama.org/en/dynamic.ch2?pageCategory.id=250

ADDITIONAL READING

Bauer, K. (2007). Wired patients: Implantable microchips and biosensors in patient care. *Cambridge Quarterly of Healthcare Ethics, 16*(3), 281–290. doi:10.1017/S0963180107070314

Farah, M. (2005). Neuroethics: the practical and the philosophical. *Trends in Cognitive Sciences, 9*(1), 34–40. doi:10.1016/j.tics.2004.12.001

Illes, J. (2003). Neuroethics in a new era of neuroimaging. *AJNR. American Journal of Neuroradiology, 24*, 1739–1741.

Chapter 8
Engineering and Environmental Technoethics

INTRODUCTION

Since modern technology was boosted by the erroneous idea that every increase of human ability to produce is to the good, and to the extent that the consequence of this idea today is a danger to the ecosystem as a whole, we need a new moral code for technological daily activities to guide both researchers and industrial organizers, if not every member of the community --Agassi, 2003, p. 251.

This chapter traces the development of Engineering Ethics, Computer Ethics, and Environmental Technoethics. It also covers the topic of military technoethics as an important new development that deserves special attention. The story begins in the

DOI: 10.4018/978-1-60566-952-6.ch008

late 19th century with the development of various engineering professional bodies to ensure that engineers were responsible for potentially harmful constructions. This in turn, gave rise to the creation of codes of engineering ethics to help guide professional conduct. As the public demand for engineering increased throughout the 20th century, so did the ethical implications and demand for codes of engineering ethics. In the 1950s and 1960s, the continued expansion of industrial growth lead also to a number of human caused environmental disasters ranging from oil spills to nuclear explosions to the release of toxic chemicals into the air and water supply. This brought on a public reaction among environmental organizations and increased public attention to ethical implications of technology and the environment. These developments helped nurture in studies in environmental technoethics and the ethical concern over human involvement in technology related environmental change. Also in the 1950s and 1960s, the public use of mainframe computers, promising outlook for computer networking, and scholarly interest in systems research raised additional interest concerning the ethical implications connected to computer innovation in society. This chapter provides a review of background developments, challenges, and current directions in each of these areas. It uses examples to illustrate the potency of technology in reference to key areas (i.e., access equity, software design, computer navigation systems, construction, mining, and other areas of technology use and misuse). It concludes with insider interviews from leading experts working in the field and recommendations on how to use technoethical inquiry to leverage the ethical use of science and technology in areas where technological innovation has created ethical challenges and dilemmas.

BACKGROUND

Engineering Ethics

During the 18th and 19th century, the industrial revolution in Great Britain was beginning to spread through Europe and North America, giving rise to major changes in agriculture, manufacturing, and transportation around the globe. In response to the growing demand for engineering expertise, multiple engineering organizations were established. In Great Britain, the Institution of Civil Engineers (established in 1818) was created to serve as a professional association for civil engineers. In the United States, the American Society of Civil Engineers (established in 1851), the American Institute of Electrical Engineers (established in 1884), the American Society of Mining Engineers (established in 1880), and the American Institute of Mining Engineers (established in 1871) helped to professionalize engineering and regulate the activities of engineers. Other important organizations in the United

States included, the National Institute for Engineering Ethics (NIEE), the National Society of Professional Engineers (NSPE), the American Institute of Chemical Engineers (AIChE), Institute of Electrical and Electronics Engineers (IEEE), and Society of Manufacturing Engineers (SME). In Canada, key associations included, the Association of Professional Engineers and Geoscientists of British Columbia (APEGBC), the Professional Engineers Ontario (PEO), and L'Ordre des ingénieurs du Québec (OIQ). These organizations (among others) helped to nurture the profession of engineering and create regulations to govern professional activities of its members.

Early work in engineering ethics grew out of efforts within engineering organizations to improve engineering practice in the early 20th century by creating standards governing engineers' responsibilities to the public and the profession. In 1912, a code of ethics was created by the American Society of Civil Engineers. According to their Fundamental Canons of the American Society of Civil Engineers (2008):

- Engineers shall hold paramount the safety, health and welfare of the public and shall strive to comply with the principles of sustainable development in the performance of their professional duties
- Engineers shall perform services only in areas of their competence
- Engineers shall issue public statements only in an objective and truthful manner
- Engineers shall act in professional matters for each employer or client as faithful agents or trustees, and shall avoid conflicts of interest
- Engineers shall build their professional reputation on the merit of their services and shall not compete unfairly with others
- Engineers shall act in such a manner as to uphold and enhance the honor, integrity, and dignity of the engineering profession and shall act with zero-tolerance for bribery, fraud, and corruption
- Engineers shall continue their professional development throughout their careers, and shall provide opportunities for the professional development of those engineers under their supervision

Also in 1912, the American Institution of Civil Engineers established a code of ethics followed by the American Society of Mining Engineers in 1914. Subsequently, other engineering associations and institutes created codes of ethics. As a result of efforts, engineering ethics became institutionalized within the professional organizations, professional licencing, and practices of engineers. By 1950s, engineering codes of ethical conduct were firmly ingrained and the majority of professional engineers working in the United States and Canada required some sort of licensing to practice. Codes of ethics set new standards of professional values and conduct in Engineering (and other major areas of technology innovation like computing and military research) which helped entrench the field of Technoethics and drive technoethical inquiry in areas of work and society affected by technological advances.

Computer Ethics

In the late 1940 and 1950s, the groundwork for Computer Ethics was pioneered by Norbert Wiener who spearheaded Cybernetics as the science of communication and control in humans and machines:

It has long been clear to me that the modern ultra-rapid computing machine was in principle an ideal central nervous system to an apparatus for automatic control; and that its input and output need not be in the form of numbers or diagrams. It might very well be, respectively, the readings of artificial sense organs, such as photoelectric cells or thermometers, and the performance of motors or solenoids ... we are already in a position to construct artificial machines of almost any degree of elaborateness of performance. (Wiener, 1948, p. 27)

Using Cybernetics, Wiener helped legitimize the social and ethical study of computers. This helped, for the first time (outside of science fiction) to bring technology into the world of human beings and the world of human beings into the technological arena. Wiener's (1954) *The Human Use of Human Beings* was an important work that explored ethical aspects of computers, setting the stage for a body of work under the heading of computer ethics.

The advent of personal computers in the 1980s brought computing into the workplace and introduced changes in how people work and interaction. Questions concerning the ethical implications of computers continued to evolve and by the mid-1980s, new work focused on conceptualizing computer ethics and offering practical ethical guidelines for computer use (Johnson, 1985; Moor, 1985). Johnson (1985) conceptualized computer ethics as a new form of ethical inquiry influenced by the advancement of computer technology which "pose new versions of standard moral problems and moral dilemmas, exacerbating the old problems, and forcing us to apply ordinary moral norms in uncharted realms" (p.1). From a slightly different perspective, Moor (1985) attempted to highlight the importance of policy in computer ethics. He states, "A central task of computer ethics is to determine what we should do in such cases, that is, formulate policies to guide our actions (Moor, 1985).

Work in computer ethics continued to grow in the 1990s through new network technologies and the popularization of the World Wide Web (WWW) allowing more computer users to be connected and in more ways than previously possible. At the same time, a newfound appreciation of the role of information was becoming increasingly important. Theoretical work in Information Ethics was applied in computer ethics (and other areas) to identify processes and interactions within the informational environment (Floridi & Sanders, 2003). Floridi and Sanders (2003) argued that information is "a necessary prerequisite for any morally responsible action to being its primary object." This broadened the meaning computer ethics to address the continued evolution and impact of computer technology in relation to information aspects and functions. This development was important and helped to distance computer ethics from an overly narrow focus on computers as the core object of computer ethics. The information ethics perspectives helped broaden conceptualizations of the object of inquiry to better map onto current work in computer networking, ubiquitous computing, and ambient technologies.

Environmental Technoethics

It can be argued that environmental ethics has been entrenched in Western and non-Western society from the time early Taoist and Stoic writings focused on appreciating and being true to nature. But it is the time period surrounding World War II that substantive work on environmental ethics expanded its focus to address artificial environments built by humans. This is largely due to the push for science and technology innovation, which was becoming increasingly treated as a measure of progress in developed societies. As will be seen in this section, it is the rise in technology related environmental changes caused by human intervention that has helped shape environmental technoethics in the 21st century.

The ethical role of science and technology within the environment has received widespread attention since World War II, particularly when dealing with multitude of technology related environmental disasters that have resulted including, industrial chemical leaks (Union Carbide gas leak, Seveso) nuclear disasters (Three Mile Island, Chernobyl), and chemical spills (Baia Mare cyanide spill) and environmental sabotage (Gulf war oil spill). In 1984, the worst industrial disaster in history occurred when a poisonous gas leak at Union Carbide India Limited (UCIL) in Bhopal resulted in the release of 15 metric tons of methyl isocyanate (MIC). It resulted from the absence of safety procedures (cut off valves) that could have prevented the contamination. This gas contaminated over 30 square miles killing over 4000 local residents instantly and 15,000 more in the years that followed (Eckerman, 2005). According to a BBC study, over 100,000 individuals continue to suffer from chronic disease subsequent to the UCIL poisonous gas exposure (BBC, 2004). In 1986, the worst nuclear power plant accident to date occurred during nuclear reactor testing at the Chernobyl Nuclear Power Plant in the Soviet Union. The release of radioactivity from the explosion contaminated populated areas in the western Soviet Union and Europe and required the resettlement of over 300,000 people. Over 50 individuals were killed by explosion with over 4,000 related cancer cases reported since the fallout (Chernobyl Forum, 2002).

From the 1990s to the 2000s, technology related disasters continued to plague the environment and affect the survival of flora and fauna. The largest oil spill in history occurred in 1991, following the invasion of Iraqi forces into Kuwait during the Gulf War. The intentional emptying of off shore oil tankers and sabotage to over 650 oil wells in Kuwait released over 1 million tons of oil into the environment. The environmental destruction contaminated the local ecosystem, killed over 20,000 seabirds, and created additional health problems the local population which lasted several years (Barzilai, Klieman, & Shidlo, 1993). In 1998, an oil pipeline at Jesse Nigeria exploded causing heavy environmental damage and death. The explosion killed over 500 individuals and led to hundreds of burn victims after the explosion created a fire which spread to nearby villages of Moosqar and Oghara. In 2001, a major cyanide spill occurred during gold mining operations in Baia Mare, Rumania. A break in a dam led to an environmental disaster when over 100 000 cubic meters of cyanide (a toxic chemical used to purify gold) contaminated water was released. Water and fish contamination lead to a drop in fish population and food poisoning felt in Rumania, Hungary and Serbia. Also, several major oil spills causing serious environmental impact have been reported throughout the 1990s and 2000s close to Rio de Janeiro, the coast of Brittany, Galapagos Isles, Milford Haven, Angola, and the coast of Spain (BBC, 2002). The continued negative impact of technology on the environment is raising public concern and creating more awareness of the broader social and ethical considerations of dealing with technologies with the power to negatively affect the environment. For this reason, ethical aspects of technology related environmental change are a priority in current technoethical inquiry.

Military Technoethics

Ethical issues surrounding the creation and use of technology for protection, destruction and invasion is as old as the military itself and can be traced back to Cicero (106BC – 43BC). St. Augustine's, *The City of God*, and St. Thomas Aquinas', *On War* are classical works in early military ethics. Notable work under just war theory provides additional insights into military ethics and ethical debates surrounding military operations and war conduct (Walzer, 1977). Within the contemporary context within which Technoethics resides, military technoethics can be most easily placed within the context of nuclear attack. In 1945, Hiroshima became the first city in history to be attacked with a nuclear weapon when it was bombed by the United States during World War II. The nuclear bomb released killed approximately 80,000 people directly and over 10,000 subsequent deaths occurred due to injury and radiation (see Eisei & Swain, 1981). This event triggered widespread concern over the use of such weaponry. Early questions in military technoethics concern the ethical and unethical use of nuclear weaponry, by whom, and under what conditions. Who should possess nuclear weapons and who should not? Under what conditions should nuclear weapons be used if at all? Do to ends justify the means? What are ethical considerations in entering continuing, and discontinuing participation in war?

TECHNOETHICS FOR ENGINEERING AND THE ENVIRONMENT

Engineering Ethics

Up until the late 1970s, the majority of work in Engineering Ethics was largely a response to the demand on Engineering organizations for professional codes of ethics, this work expanded in depth and scope. In the 1980s and 1990s, Engineering Ethics began to become entrenched in the curriculum of many university degree programs in Engineering and a focus in engineering teaching research. This helped expand the scope of Engineering Ethics to address value dimensions of engineering as revealed in program studies, public opinion research, and media coverage. Johnson's (1991) *Ethical Issues in Engineering* and Unger's (1982) *Controlling Technology: Ethics and the Responsible Engineer* exemplify the expanded scope of technoethical inquiry into the world of engineering as an important (and potentially dangerous) social enterprise within society that requires knowledge from a diversity of contexts and expertise derived from the input of multiple perspectives and stakeholder groups. Also in 1980s and 1990s, new technological options became available which promised to reduce environmental destruction created by older technologies. This was largely in response to oil shortages in the 1970s and fear of global fossil fuel depletion. The production of wind powered electrical energy, solar panels, and the production of fossil fuel substitutes (geothermal, methanol, ethanol, etc.) is promising but has yet to surpass conventional petroleum as a cheap and available fuel source. The availability of new technologies offers new options for improving the ethical use of environmental resources.

Martin and Schinzinger's (2005)*Ethics in Engineering* provides a particularly insightful introduction to contempory engineering ethics which reinforces early work in Technoethics and arguments for greater ethical responsibility in the training and practice of engineers and technologists. A number of suggestions are offered by Martin and Schinzinger (2005):

- Engineers shall act in professional matters for each employer or client as faithful agents or trustees, and shall avoid conflicts of interest

- Engineers shall hold paramount the safety, health, and welfare of the public in the performance of their professional duties
- Engineers shall act in such a manner as to uphold and enhance the honor, integrity, and dignity of the engineering profession
- Engineers shall continue their professional development throughout their careers, and shall provide opportunities for the professional development of those engineers under their supervision
- Engineers shall continue their professional development throughout their careers, and shall provide opportunities for the professional development of those engineers under their supervision
- Engineers should keep current in their specialty fields by engaging in professional practice, participating in continuing education courses, reading in the technical literature, and attending professional meetings...'(p. 308-312)

The broadening scope of ethical inquiry into engineering maintains a deep concern for professional responsibility issues from earlier work while expanding the realm of engineering ethics to include public needs and interests. Typical questions in technoethical inquiry include, "To what extent should engineers and other stakeholders be responsible for the negative impacts of Engineering on society?", "To what extent and under what circumstances should the public be informed of engineering risks?", "How should society deal with risky Engineering applications which could potentially be used for evil? What are the ethical considerations in failing to adopt available technology solutions that are more publically acceptable?

Computer Ethics

Computer Ethics is a branch of Technoethics concerned with the human use of computer technology in a number of areas, including computing in the workplace, graphic interfaces, visual technology, artificial intelligence, and robotics. Over the last 10 years in particular, interest in computer ethics has expanded to include complex systems, robotics, and questions concerning the moral agency in artificial intelligence. One major challenge in current technoethical inquiry focuses on whether artificial agents should also be considered artificial moral agents (AMA) worthy of moral concern. Research on autonomous agency has become a new focus in computer ethics, partly due to the advent of innovative work in artificial intelligence (AI). Artificial intelligence (AI) contributes robotic technologies with new capabilities previously possible only in humans, thus raising legal and ethical implications concerning the possibility of autonomous agency in non-human entities. Franklin and Graesser (1996) define an autonomous agent, as "a system situated within and a part of an environment that senses that environment and acts on it, over time, in pursuit of its own agenda and so as to effect what it senses in the future." Advancing technologies present new human and technology relationships which are challenging traditional notion of moral agency. Sullins (2008) notes that artificial agents "present unique challenges to the traditional notion of moral agency and that any successful technoethics must seriously consider that these artificial agents may indeed be artificial moral agents (AMA), worthy of moral concern."

Current work in computer ethics continues to address core questions about computing technology while taking on new questions that have emerged with the rise of increasingly sophisticated technologies. Because computer technology is continually advancing, there is an ongoing need for developers, users, and other stakeholders to address ethical implications that affect individuals in a variety of ways. What are the new ethical implications for computer security and privacy in post-9/11 society? What are the ethical considerations in relations between software development companies, customers, competitors, and the market? What ethical considerations apply to computer hacking and viruses? What ethical considerations are there concerning computer access and control? How do stakeholders reach consensus concerning new developments in computing that have potential for harm and good? To what extant can sophisticated artificial agency be attributed ethical rights and responsibilities? To what extent are ethical responsibilities in artificial agents similar to that of human agents?

Environmental Technoethics

Attfield's (2003) *Environmental Ethics: An Overview for the Twentieth Century* provides a recent review of current environmental problems with an environmental ethics focus on sustainable development. The text highlights the importance of adapting a precautionary principle in the face of uncertainty, particularly when, "there is reason to assume that certain damage or harmful effects on the living resources of the sea are likely to be caused by such substances, even when there is no scientific evidence to prove a causal link between emissions and effects" (p. 144). The types of environmental problems faced within the knowledge society are largely human induced and technologically mediated, which must be nderstood to be addressed. In a similar vein, Robert Traer's (2009)*Doing Environmental Ethics* highlights the role of technology within our current ecological crisis along with the need to cultivate a deeper sense of self and one's place in nature in order to reduce the negative effects of an increasingly damaging human footprint mediated by powerful technologies. One major lesson learned from this text is how learning from nature requires an attention to consumption habits, ecological living, and the creation of environmental policies from multiple stakeholder groups (I.e. governments, corporations, Non-governmental organizations).

Current work in environmental technoethics examines ethical aspects of environmental change due to manufacturing, mining, and industry. In recent years, environmental technoethics has began to address other technologically infused environments including housing projects, urban planning, urban living, and slums. This broadened focus of environmental technoethics on technologically infused environments has contributed in important ways. First, it appeals to a larger audience than ever before. The new ethical dilemmas of urban living concerns all members of society, not only environmentalists. Second, the broader scope of environmental technoethics has opened up technoethical inquiry to consider macro level issues that are receiving attention in disciplines outside environmental studies such as Sociology and Economics. The reason for this is that environmental technoethics provides a broad reaching focus on important macro level issues connected to the influence of artificial objects on natural environments including air pollution, environmental destruction, globalization effects, unequal resource consumption, and poverty. Amassing work in environmental technoethics continues to expand the boundaries of this interdisciplinary area of research concerned with artificial environments and the relation of constructed environments (technology) to natural environments, human life, and other living organisms (King 2000, Light 2001). Environmental technoethics brings together work from environment scientists, geographers, economists, political scientists are others working at the crossroads of technology and environmental ethics. One area of popular inquiry in environmental technoethics focuses on globalization and the "wicked" problems connected to it including, global inequality, unequal resource access, poverty, unequal opportunities, unequal wealth distribution, and other underlying mechanisms sustaining inequality. Some of these have well documented by documented by leading sociologists like Charles Tilly (Tilly, 1998). In the area of Economics, relations between technology, economics and environmental issues have been researched in an effort to demonstrate the complicated network of factors that can lead to technology driven problems with serious ethical implications for society like environmental destruction (Shrader-Frechette, 1984) and urban decay (Light, 2001).

Environmental Technoethics provides an interdisciplinary research focus based on achieving a balanced inquiry and multiple perspectives. On the one hand, the creation of artificial objects and constructed environments is critical to the economy, military protection services, education, and leisure. On the other hand, overreliance on technology is also responsible for a great deal of environmental destruction and pollution. For example, a significant proportion of citizens in developed countries like the United States are able to afford cars which is a key contributor to air pollution in urban settings. If the majority of the population in developing countries also had access to cars, the level of air pollution would quickly choke out human life on our planet. To this end, environmental technoethics delves into complex problems requiring interdisciplinary collaboration among philosophers and theorists in the social as well as the natural sciences in order to have a balanced inquiry that focuses on both positive and negative technology relations from a variety of perspectives.

The growth of environmental technoethics has evolved through interdisciplinary synergies emerging from a variety of areas where the use of technology leads to the creation of artificial environments and changes within natural environments caused by human driven technological influence. This has also contributed new research questions to help guide research in environmental technoetrhics. How can authorities deal with slums and poverty within urban settings in developing countries? What are the ethical considerations in urban pollution? Who is ethically entitled to secure and exploit limited natural resources like oil? What are the ethical guidelines for urban planning? In an increasingly crowded industrialized world, the answers to such problems are pressing.

Military Technoethics

Military Technoethics is framed around the study of ethical aspects of military operations involving technology. The use of advanced technologies has become increasingly important in military operations which has sparked a great deal of public attention and debate concerning the ethical use of technology in the military. This is reflected in recent media attention to ethical aspects of technology use in war and terrorism in post 9/11. Steinhoff's (2007) *On the Ethics of War and Terrorism* is an excellent example of the resulting new scholarship in military technoethics that seeks to resituate the discussion of just war theory within the current context of military operations where technology can have a global impact, "For the survival and freedom of political communities — whose members share a way of life, developed by their ancestors, to be passed on to their children — are the highest values of international society" (133). New work in military technoethics raises a number of new ethical questions regarding the appropriate use of advanced technologies in situations where the military must act. Should soldiers and governments use technology to overrule the rights of innocent people for the sake of their own community? What are the new ethical considerations in the creation and use of new military technologies? Who is ethically entitled to create and use weapons of mass destruction? What are the ethical guidelines for using high risk technologies to begin or end a war?

TECHNOETHICAL FRAMING

Based on the reviewed grounding work about technoethics for computing, engineering and the environment, it is clear that technology has changed the world and the role of humans within it. This is particularly salient for technology professionals who are on the front lines responsible for designing technological innovations that impact the world. In this new world of technology, technology professionals are viewed as both innovation leaders and ethical agents with a responsibility to the world. Professionals working with technology not only have to address design and production issues, they also must deal with a plethora of ethical considerations connected to their work.

A general approach to technoethical inquiry builds on Bunge's (1977) recommendations to situate technoethical decision making in a larger framework, "Because of the close relationships among the physical, biological and social aspects of any large-scale technological project, advanced large-scale technology cannot be one-sided, in the service of narrow interests, short-sighted, and beyond control: it is many-sided, socially oriented, farsighted, and morally bridled" (Bunge, 1977, 101). When working on any technology initiative, there are interests and needs from a variety of sources to consider: the organization desires an efficient plan, the organization members desire good working conditions, citizens desire a clean and safe environment and useful products and services that are reasonably priced. The types of decisions made depend on the weighing of technoethical knowledge derived from key relational dimensions.

An examination of current oil and gas production technology provides a useful example of how technoethical inquiry is applied. There is a growing controversy over the use of technology to extract non-conventional oil known as oil sands (or tar sands) from the environment. Given the high demand and limited reserves of crude oil worldwide, oil sands represent a major part of the oil industry today despite public concern and environmental risks involved in its production. Oil sands are a naturally occurring mixture of sand, water, and bitumen (a form of petroleum) used to make synthetic oil. Canada has the largest quantity of oil sands in the world and is the largest supplier internationally. It is estimated that Canadian oil sands represent over 60% of the world's total petroleum resource with 173 to 315 billion barrels of oil in the oil sands potentially recoverable with today's technology. A second major controversy surrounds the extraction of sour gas from the environment, a natural gas containing high levels (over 1%) of hydrogen sulfide (H2S). Approximately one third of Alberta's natural gas is sour gas and Alberta is responsible for 85% of Canada's sour gas production, the majority of which is located in the eastern foothills of the Rocky Mountains (CBC, 2005). Hydrogen sulfide is a toxic gas that can cause headaches, loss of consciousness, and death. Although most recorded deaths have been workers, the lack of research on long term effects of hydrogen sulfide to sustained low-level exposure by people living near sour gas wells has raised serious public concern and led to illegal oil pipeline bombings near Dawson, located close to the British Columbia and Alberta.

At the environmental level, the challenge of protecting the environment during development of this resource has become a priority for the Alberta government. More stringent legislation and on-the-ground measures are already in place to protect the air, land and water during oil sands and sour gas development. At a societal level, the Alberta government has established an Oil Sands Environmental Management Division in Alberta Environment to safeguard the environment during oil sands and sour gas development. At the organizational level, the Oil Sands Developers Group was established in 2008 (originally called the Regional Infrastructure Working Group) to deal with issues related to the development of resources within the Athabasca Oil Sands Deposit region of Alberta. The group works closely with oil sands developers, government and other stakeholders affected by the oil industry. The group examines available information and develops strategies to resolve issues using the following processes:

- All issues are reviewed for completeness and understanding
- A priority is assigned
- Relevant stakeholders are invited to address the issue
- The scope of work is reviewed with input from new stakeholders
- The issue is reviewed for completeness, availability of data and shortfalls identified
- Studies may be carried out if data is insufficient
- Information is collated and passed on to the appropriate regulatory body
- Progress and findings are reported to The Oil Sands Developers Group committee and Board of Directors on a continuous basis

Although the citizen level does not have a strong influence, the possibility of citizen participation in oil sand development is promising and could be easily incorporated into the other levels of technoethical consideration. Despite this area to improve, oil sand production appears to have a moderate level of technoethical knowledge with some room for improvement. The same does not hold true of sour gas development which lacks a similar structure to address environmental, societal, citizen, and organizational levels of technoethical consideration. This could partly be attributed to the gap in understanding and experience with oil sand and sour gas production on the British Columbia side of the provincial border. The lack of technoethical consideration in sour gas production at the citizen level has led to vandalism, sabotage of the gas pipelines, and considerable revenue loss (CBC, October 16, 2008). Sour gas production appears to have a low technoethical knowledge and a great deal of improvement to make.

Overall, this rough sketch of a technoethical inquiry is consistent with existing models of ethical analysis within individual branches of technoethics. Floridi's (2003) model of moral action accommodates similarly broad considerations including, the moral agent, the moral client, the interactions of these agents, the agent's frame of information, the factual information available to the agent, and the environment where the interaction occurs.

DISCUSSION QUESTIONS

- How do you think military research involving nanoscale technologies and neuroscience should be regulated?
- Are government efforts to create weapons of mass destruction justified in the name of national security?
- Imagine you were part of an evaluation board deciding whether or not to give financial support to a new technology innovation project with a high level of environmental risk? What ethical considerations would you address (if any)?

DISCUSSION

Mario Bunge (1977) points out the immense influence that the scientists, technologists, engineers and managers have on our society. In examining the controversy over oil and gas development in Alberta and surrounding areas, it was found that oil sand development in Alberta had a fairly strong relational network of technoethical knowledge to ground current operations and future planning. It was also found that sour gas development had a fairly weak relational network of technoethical knowledge sharing to ground current operations and future planning. Drawing on rich context-specific examples (like the oil and sour gas case) help to explain how serious consequences (i.e., illegal pipeline bombing) could arise in response to a perceived weakness in technoethical knowledge linking key areas of concern within environmental, societal, citizen, and organizational areas. Therefore, the grounding of technoethical expertise and knowledge requires the focus on real life context-specific examples of human-technology relations including, participation from institutionalized multi-stakeholder organizations responsible for addressing environmental issues and concerns, technoethical knowledge building about factors affecting societal acceptance of controversial technology development and application, and knowledge about factors affecting citizen acceptance of controversial technology development and application. Building on the emphasis on human-technology relations, an interview with leading scholar, Dr. A. Pablo Iannone provides a promising outlook for continued work in technoethical inquiry based on trancultural and transnational dialogue building to address major technoethical problems that have a global impact.

REFERENCES

American Society of Civil Engineers. (2008). *Code of ethics*. Reston, VA: ASCE Press. Retrieved October 15, 2008, from http://www.asce.org/inside/codeofethics. cfm

Attfield, R. (2003). *Environmental ethics: An overview for the twentieth century*. Cambridge, UK: Polity Press.

Barzilai, G., Klieman, A., & Shidlo, G. (Eds.). (1993). *The Gulf crisis and its global aftermath*. New York: Routledge.

BBC. (November, 2002). *Comparing the worst oil spills*. Retrieved October 31, 2008 from, http://news.bbc.co.uk/2/hi/europe/2491317.stm

BBC. (November 30, 2004). *Bhopal faces risk of 'poisoning*. *BBC Radio 5*. Retrieved October 31, 2008 from http://search.bbc.co.uk/cgibin/search/results.pl?scope=all& tab=av&recipe=all&q=bhopal+faces+risk+of+%27poisoning%27&x=0&y=0

Bunge, M. (1977). Towards a Technoethics. *The Monist, 60*(1), 96–107.

CBC. (August 10, 2005). *Sour gas: Why it makes people nervous*. Retrieved October 29, 2008, from http://www.cbc.ca/news/background/environment/sour_gas.html

CBC. (October 16, 2008). *More sabotage feared after 2nd pipeline bombed in northern B.C.* Retrieved October 29, 2008, from http://www.cbc.ca/money/story/2008/10/16/ bc-second-pipeline-explosion-dawson-creek.html

Chernobyl Forum. (2002). Chernobyl: Assessment of radiological and health impact. Nuclear Energy Agency. Retrieved October 31, 2008 from http://www.nea.fr/html/ rp/chernobyl/c01.html

Eckerman, I. (2005). The Bhopal gas leak: Analyses of causes and consequences by three different models. *Journal of Loss Prevention in the Process Industries, 18*, 213–217. doi:10.1016/j.jlp.2005.07.007

Eisei, I., & Swain, D. (1981). *Hiroshima and Nagasaki: The physical, medical, and social effects of the atomic bombings*. New York: Basic Books.

Floridi, L., & Sanders, J. (2003). Computer ethics: Mapping the foundationalist debate. *Ethics and Information Technology, 4*(1), 1–24. doi:10.1023/A:1015209807065

Franklin, S., & Graesser, A. (1996). Is it an agent, or just a program: A taxonomy for autonomous agents. In *Proceedings of the Third International Workshop on Agent Theories, Architectures, and Languages*. New York: Springer-Verlag.

Johnson, D. (1991), *Ethical issues in engineering.* Englewood Cliffs, NJ: Prentice-Hall.

Karliner, J. (1997). *The corporate planet.* San Francisco: Sierra Club Books.

King, R. (2000). Environmental ethics and the built environment. *Environmental Ethics, 22,* 115–131.

Light, A. (2001). The urban blindspot in environmental ethics. *Environmental Politics, 10,* 7–35.

Martin, M. W., & Schinzinger, R. (2005). *Ethics in engineering.* Boston: McGraw-Hill.

Moor, J. H. (1985). What is computer ethics. In T. W. Bynum (Ed.), *Computers and ethics* (pp. 266-275). Oxford, UK: Basil Blackwell.

Moor, J. H. (2005). Why we need better ethics for emerging technologies. *Ethics and Information Technology, 7,* 111–119. doi:10.1007/s10676-006-0008-0

Oil Sands Developers Group. (2008). Retrieved October 29, 2008, from http://www.oilsands.cc/issues/

Shrader-Frechette, K. (1984). *Science policy, ethics and economic methodology.* Dordrecht: D Reidel.

Steinhoff, U. (2007). *On the ethics of war and terrorism.* Oxford: Oxford University Press.

Sullins, J. (2008). Artificial moral agency in technoethics. In R. Luppicini & R. Adell (eds), *Handbook of research on technoethics* (pp. 205-221). Hershey: Idea Group.

Tilly, C. (1998). *Durable inquality.* San Francisco: University of California Press.

Traer, R. (2009). *Doing environmental ethics.* New York: Westview Press.

Unger, S. (1982). *Controlling technology: Ethics and the responsible engineer.* New York: Holt, Rinehart and Winston.

Walzer, M. (1977). *Just and unjust wars: A moral argument with historical illustrations.* New York: Basic Books.

Wiener, N. (1948). *Cybernetics: Or control and communication in the animal and the machine.* Cambridge, UK: The Technology Press.

Wiener, N. (1954). *Human use of human beings,* (2nd Ed). New York: Houghton Mifflin

ADDITIONAL READING

Attfield, R. (2003). *Environmental ethics: An overview for the twentieth century.* Cambridge: Polity Press.

Cheney, G. (1995). *Journey to Chernobyl: Encounters in a radioactive zone.* Chicago: Academy Chicago.

Curry, P. (2005). *Ecological ethics: An introduction.* Cambridge: Polity Press.

Kostrewski, B. J., & Oppenheim, C. (1980). Ethics in information science. *Journal of Information Science, 1*(5), 277–283. doi:10.1177/016555157900100505

Latour, B. (2004). *Politics of nature: How to bring the sciences into democracy.* Cambridge, MA: Harvard University Press.

Light, A., & Rolston, H. (Eds.). (2003). *Environmental ethics: An anthology.* Oxford: Blackwell.

Martin, M., & Schinzinger, R. (2005). *Ethics in engineering.* Boston: McGraw-Hill.

Chapter 9
Educational and Professional Technoethics

INTRODUCTION

Are we developing a (global) society where our youth think it is ok to copy and paste whatever they see on the Internet and turn it in for homework; where writing an English paper would include BTW, IMHO, LOL among other emoticons; where downloading a song or movie that they can pirate from the Web is perfectly ok? We would certainly hope not. However, these concerns are just the tip of what is happening in our society. When looking at the social impact of technology in on our society it becomes clear the importance of instilling ethical behaviors and practices in the members of our society. Where is the best place to instill these ethical behaviors? --Gearhart, 2008, p.263.

DOI: 10.4018/978-1-60566-952-6.ch009

The above passage provides a glimpse at the challenges connected to education and the increasing role of technology within it. As a formal field of study, Educational technology is pursued a variety of research programs under various names which focus on the connections between technology and education within society (i.e., Instructional Technology, Educational Computing, Distance Education, Technology Education). This work places technology oriented developments in education under the general heading of educational technology. As defined by Luppicini (2005):

Educational Technology is a goal oriented problem-solving systems approach utilizing tools, techniques, theories, and methods from multiple knowledge domains, to: (1) design, develop, and evaluate, human and mechanical resources efficiently and effectively in order to facilitate and leverage all aspects of learning, and (2) guide change agency and transformation of educational systems and practices in order to contribute to influencing change in society (p.108).

Educational technology deals with the use of technology within traditional educational settings (i.e., schools, colleges, universities) as well as professional training contexts where technology is influencing work and professional activities (i.e., technical schools, public institutions, companies, non-governmental organizations).

Ethical considerations within amassing work in educational technology refocuses attention from technology and education to ethics, technology and education. Muffoletto (2003) states the "concept of ethics within the field of educational technology is a conceptual construct that legitimates certain behaviors and interpretations, while reifying particular social structures and ways of knowing" (p. 62). Educational technoethics is dedicated to social and ethical aspects of technology within formal and informal educational contexts (education, training, and evaluation). Thus, it is divided into two main areas dealing with traditional educational settings as well as professional educational contexts. General educational technoethics is concerned with ethical use of technology to promote the aims of education while professional technoethics is a specialized area of educational technoethics focusing on the development and evaluation of ethical codes and standards to guide decision-making about technology in education and professional life. Within educational technoethics, this often involves the strategic use of technology assessment models to evaluate new technologies and their possible consequences, along with the assignment of social and ethical responsibility. This chapter traces the roots of technoethical inquiry in education and professional life. It also explores current efforts in educational technoethics to leverage the use of technology in education and professional life.

BACKGROUND

In the early 1970s, early work in educational technoethics was carried out in the field of Educational Technology through the institutionalization of educational technology codes of ethics and later academic work from individuals like Nichols (1987) and Boyd (1991). Nichols (1987) discussed the positive and negative aspects of educational technology in an effort to derive moral guidelines for educational use. Boyd (1991) conducted early work in educational technoethics framed within a program of emancipatory educational technology taken to be a means to foster positive educational values using new technology. The motivation driving this early work was to help raise awareness about the dangers of technology in education. It was also intended to encourage an ethical examination of technology as a necessary condition for appropriate technology use in education. In the mid-1990s, educational technethics achieved widespread recognition with the popularization of the Internet and the World Wide Web (WWW) which intensified problems connected to privacy violations, breaches in intellectual property, student harassment, and cheating. The popularization of the WWW gave students easier access and more opportunities to cut and paste from others' work or purchase others' work online.

Juxtaposed with educational technoethics was specialized work under the umbrella of professional technoethics. This work first evolved under Bunge (1977), who pioneered technoethics as a professional obligation directed at engineers and other technology workers. This work took an important step in pushing for greater accountability among technology professions over their creations and raising awareness about the social and ethical responsibilities connected to technology and professional life. As an important area of educational technoethics, professional technoethics deals with the issues of ethical responsibility for professionals working with technology such as, engineers, scientists, computer and software designers, military personnel, medical researchers, etc.

Developments in educational technoethics within education and professional life raised public awareness about the need to critically evaluate technologies. This gave rise to various technology assessment models and strategies for safeguarding the public from potentially harmful technological applications. In 1972, the Office of Technology Assessment (OTA) was established within the United States Congress to provide expert analyses documenting complex scientific and technical issues (OTA Publications, 2008). This organization raised awareness about the importance of evaluating new technologies which served as a model for other countries concerned with the consequences of technologies on society including Denmark (Danish Technology Board), Great Britain (Parliamentary Office of Science and Technology), Netherlands (Flemish Institute for Science and Technology Assessment) and Germany (Institute of Technology Assessment and Systems Analysis). In 1990, the European Parliamentary Technology Assessment (EPTA) was created to bring together member organizations from across Europe. The mandate of EPTA is to "advance the establishment of technology assessment as an integral part of policy consulting in parliamentary decision-making processes in Europe, and to strengthen the links between technology assessment units in Europe (European Parliamentary Technology Assessment, 2008). Although most technology assessment programs have not focused solely on ethics, most technological assessment strategies have an essential ethical component to provide benefits and balance benefits against risks (beneficence), to avoid causing harm (non-malfeasance), to respect the decision-making capacities of others (respect), and to be fair in the distribution of benefits and risks (justice). The continued expansion of technoethical considerations in education, professional life, and evaluation work helped set the stage for follow-up work in these continued areas of technology interest.

CURRENT WORK IN EDUCATIONAL TECHNOETHICS

Digital Divide and Educational Access

The digital divide in education takes on a number of dimensions. First, there is the digital divide in Internet access which refers to the gap between those you have access to the Internet and those that do not. Despite the rapid development of broadband saturation in countries like Canada and Korea, other counties do not have the financial resources to create the same Internet infrastructure. This leads to gaps in educational content quality between those able to access and share a breadth of educational resources (text, images, audio, film) and those with limited capacities (text only). What is more alarming is the gap between regions and people that have Internet access from those that have no access at all. As described by Ryder (2008):

This expression [digital divide] arose in the digital age to describe the information gulf that exists between peoples and societies. The perceived gulf is the result of the dramatic rise of information technologies that evolved exponentially in the developed countries during the latter half of the Twentieth Century. The expression connotes the idea that information is a potent source of power, and those who enjoy access to information technologies have the potential to wield significant power over those who have no such access. In particular, it denotes the manner that technologies undergo a metamorphosis through the process of adoption and use over time (p.247).

What is important to note is the complexity of the digital divide in information that goes beyond technical considerations or designer intent. The digital divide is also affected by political forces and human values which also affect the digital divide in information and how it is dealt with.

Second, there is also the digital technology divide which refers to the gap between those who are able to use digital technology and those are not. For instance, there are schools and students who have the latest digital technologies (i.e., iPhone, PDAs, Blackberry, Laptops, etc.) and others who have outdated equipment or none. While Universities in North America like Acadia, Duke, and Wakefield have been able to provide each student with laptop computers, access to electronic technologies at other instituations is limited due to costs. Depending on budget constraints and priorities, many schools (and students) are left behind within an age defined by technology and progress. These digital divides in education describe the pervasive gap in level of digital technologies and technology related services (Internet access) between schools in both developed and developing countries. This gap leads to inequalities in life opportunities since groups in society with higher levels of access to digital technology will have higher technological literacy and increased opportunities in for higher paying careers within the technologically driven knowledge economy.

Educational technoethics also deals with projects intended to combat the digital divide such as the OLPC (One Laptop Per Child) project established in 2005 by Nicholas Negroponte within the M.I.T. Media Lab. The OLPC project is an aggressive project attempted to reduce the digital divide by developing a $100 laptop which could be distributed to children in developing nations around the work (MIT Media Lab, 2009). Ryder (2008) provides a technoethical analysis of the XO laptop design with particular interest on the ethical territory that is traversed in the implementation of a low-cost computer intended for millions of children in underdeveloped nations. Ryder (2008) noted "how XO designers negotiated between ethics and feasibility as they confronted numerous problems including infrastructure, power consumption, hazardous materials, free vs. proprietary software, security, and the cost of labor (p.232). Technoethical analyses like this go beyond technical considerations and allow social and ethical factors to be considered, such as how the imposition of a new educational technology needs to address issues such as competing stakeholder interests, technology control, user needs, socio-cultural values, and relative importance.

Information Privacy Issues and Censorship in Schools

With the increased capacity for Internet information access and sharing comes the burden of controlling this information and protecting its users. This is particularly salient in education. Access to inappropriate content (e.g., pornography, pirated music and films, online violence, gambling sites, etc) is a major concern for teachers, parents and school administrators. Because computers and Internet services are typically owned by the educational institution, censorship is used to control Internet access. There are a number of filtering softwares (e.g., Child Safe, Cyber Patrol, Net Nanny, etc.) which along restrictions to be placed on what is viewed on site. However, this does not block student access using their own internet service providers at school. Neither does it acknowledge that many students are able to circumnavigate the best censorship controls quickly after they are implemented.

A related concern is about information privacy issues. The medium of the Internet allows information to be copied and transferred easily via email systems which have limited security. Despite censorship measures taken in many educational institutions, there remains problem of misusing the Internet and Internet applications like social networking tools (e.g., friendster, facebook, you tube, myspace, etc.) to violate the privacy of others as is the case in cyber-bullying, online harassment, and information theft. Given the copy and paste functionality of WWW, it is possible to post private messages, images, or film clips taken from individual without their consent. Information related issues in Internet use have led to a variety of technoethical studies on cyberloafing (Lim, 2002), network attacks (Wilson, 2005), information theft, unauthorized transmission of confidential information (Redding, 1996), transmission of offensive jokes, deceptive communication, online flaming (Gurak, 2001), and online identity deception (Donath, 1999).

Plagiarism and Cheating

Internet plagiarism (Lathrop & Foss, 2000) and related academic offences (Underwood & Szabo, 2003) is one of the most widely discussed areas of educational technoethics. The capacity for copying and pasting digital images and text from the WWW has led to public debate and led to the creation of new copyright legislation such as the US Digital Millennium Copyright Act, and the European Union Copyright Directive. This legislation is intended to protect intellectual property of authored works digitally represented and transmitted over the WWW. Within educational institutions (schools, colleges, university), copyright rules are typically connected to plagiarism issues (students copying texts or images from the Web and using it without citing its source or passing it off as one's own). A number of studies have shown that academic dishonesty is a serious ethical dilemma in educational institutions (Ashworth, Bannister, & Thorne 1997; Lathrop & Foss, 2000). In response, private companies have partnered with a number of educational institutions in the creation of Internet applications like Turnitin.com to detect plagiarism (Turnitin, 2007). Sites like Turnitin.com contain huge database of published and unpublished authored work. When a paper is submitted, the service compares textual patterns in the submitted paper to the database contents. When submitted papers contain the same or a similar pattern of words, the user (instructor, teaching assistant, etc) is notified so that the paper can be thoroughly reviewed to determine if the student has copied from the source. Such Internet services act as a deterrent against copying and pasting digitally represented text and images without permission. However, countermeasures like this are controversial since student material uploaded to many of these online plagiarism detection sites is copied and retained without student permission.

There are also other technoethical dilemma's connected to new technology such as cheating. Although cheating has existed as long as educational institutions, new technologies (e.g., mobile phones, mini cameras, PDA's, digital microphones, etc.) help facilitate cheating by allowing students to record and transmit test paper questions and answers to other students (Socol, 2006). This situation has created serious technoethical challenges for educators and administrators to deal with. As stated by Pullen (2008):

As the technology develops and students get better at using the latest and smallest devices, teachers and school officials find it harder to cope with cheaters. The more cheating goes uncaught, the freer students feel to do it. This is a vicious cycle. Ethically and professionally, schools have a duty to prevent cheating which must go hand in hand with clear standards and policies about what cheating is (p.687).

Technoethical Teaching and Training

Technoethical inquiry is also studied in a variety of teaching and learning contexts outside general education. For instance, the application of assistive technologies in special education contexts is a challenging area of educational technoethics. Students with mental and physical disabilities often need specialized technologies and services that strain limited educational budgets that have to support special education and general education. This can lead to inequalities in educational services depending on the priorities of the educational institution. In response, new approaches educational administrations are exploring solutions to such ethical dilemmas. A case study by Candor (2008) explores how one school district responds to this ethical dilemma by showing how special education technology can benefit both for general and special education needs. Candor (2008) argues that the "allocation of resources for assistive technology does not have to result in a gap between general and special education"(p.411). This, however, creates new challenges for educational institutions such as the adaptation of assistive technology to the curriculum. As indicated by Pullen (2008):

This will not usually be easy. Students with visual impairment will find many web-sites hard to read, though screen reader programs can help if the webpage has been designed for such use. Web designers generally design webpages for older browsers but not always for adaptive devices (p. 694)

There is a need for teachers to integrate educational technoethics in their teaching. Pullen (2008) argues, "Teachers need to understand the social, ethical, legal and human issues that surround the use of digital technology in schools and then apply that understanding to help students become technologically responsible" (p. 680). One way this can be done is by focusing on technology and ethics within the curriculum. For instance, Tavani (2007) suggests the following approach for teachers attempting to integrate technoethics into their teaching:

- **Step 1:** Identify a practice involving technology or a feature in that technology that is controversial from a moral perspective
  - 1a. Disclose any hidden features or issues that have moral implications
  - 1b. If the issue is descriptive, assess the sociological implications for relevant social institutions and socio-demographic groups
  - 1c. If there are no ethical/normative issues, stop
  - 1d. If the ethical issue is professional in nature, assess it in terms of existing codes of conduct/ethics for relevant professional associations
  - 1e. If one or more ethical issues remain, go to step 2
- **Step 2:** Analyze the ethical issue by clarifying concepts and situating it in a context
  - 2a. If a policy vacuum exists, go to step 2b; otherwise go to step 3
  - 2b. Clear up any conceptual muddles involving the policy vacuum and go to step 3
- **Step 3:** Deliberate on the ethical issue. The deliberation process requires 2 stages
  - 3a. Apply one or more ethical theories to the analysis of the moral issue, and then go to step 3b
  - 3b. Justify the position you reached by evaluating it against the rules of logic/critical thinking (Pgs. 23-24)

In the professional training context, there is an urgent need for technology focused curriculum with an ethical and social orientation. Haghi, Mottaghitalab, and Akbari (2008) draw on over 20 years experience teaching engineering to discuss the changes in pedagogical training for engineering faculty in order to better prepare engineering students. Because of the close relationship to technology in the field of engineering, a high level of attention is directed at engineering and how to teach it to prepare future engineers to be socially and ethically responsible in their work. Haghi, Mottaghitalab, and Akbari (2008) note, "Increasing emphasis is being placed on establishing teaching and learning centers at the institutional level with the stated objective of improving the quality of teaching and education" (p.439). In a different area, Robbins, Fleischmann, and Wallace (2008) address the need to develop new approaches to teaching computing and information ethics (CIE). This article explore three domains where information ethics may be applied to education: Information ownership; information privacy; and information quality. Robbins, Fleischmann, and Wallace (2008) suggest the need to," focus upon developing a deep understanding of the relationships between students, teachers, pedagogical materials, learning processes, teaching techniques, outcomes and assessment methods (p. 391). The changing landscape of teaching in technology oriented fields is helping to infuse teaching with technoethical principles to better serve the public.

Professional Technoethics

Current work in professional technoethics focuses on standards of practice and codes of ethical conduct to help professionals work responsibly with technology. This includes anyone involved in the design and development of technology as well as other professionals working with technology in their day-to-day activities. Professional technoethics views ethical decisions made during the development and use of technology as important in the advancement of a technological society which satisfies the interests of its members. Work is underway to establish useful codes of ethics and licensing standards for professionals including, the ACM Code of Ethics and Professional Conduct and professional licensing standards for software engineers. This is done in an effort to ensure a high level of professional responsibility and accountability at the core of technology innovation.

One of the reasons for focusing on ethical standards and codes is to leverage professional responsibility in areas of technology development is that these the professionals working with technology have specialized knowledge, expertise, and respect as technology authorities in the public eye. This provides technology professionals with opportunities to play an important role in how technology is used in the world. This follows the age old credo, "with great power comes great responsibility". Key technology oriented professional organizations (E.g. Association for Computing Machinery (ACM), Institute of Electrical and Electronic Engineers (IEEE)) have well articulated codes of ethics and accreditation requirements to guide technology professionals in fulfilling their ethical responsibilities. The ACM Code of Ethics highlights core moral imperatives for computer professionals such as, avoiding harm to other, being honest and trustworthy, and respecting existing laws pertaining to professional work. In a similar vein, the IEEE Code of Ethics emphasizes the avoidance of real or perceived conflicts of interest whenever possible and the need to be honest and realistic in stating claims or predictions based on available data. In addition, a number of accreditation boards work to enforce ethical standards in teaching and training within varies technology oriented professions. The Accreditation Board for Engineering Technologies (ABET) the the Computer Sciences Accreditation Commission/Computer Sciences Accreditation Board (CSAC/CSAB) and requires an ethics training in computer engineering curriculum. Thus, ethical standards and codes for technology professionals are an important part of contemporary professional life for technology training and practice.

DEFINING A TECHNOETHICAL FRAMEWORK FOR EDUCATION AND WORK

The participatory turn in technology assessment has been popularized in the last 10 years partly in response to overly narrow (and often biased) approaches to expert led technology assessments. This is particularly salient in international development work where technology initiatives are implemented in an effort to aid an entire community with unique interests and needs that have to be taken into consideration. Technology professionals are challenged by a variety of professional relationships with a network of technology stakeholders [employer—technology professional, client—technology professional, professional—professional, society-- technology professional). This requires an awareness of a diversity of interests, which are important to take into consideration in order to satisfy all involved. Guba and Lincoln's (1989) *Fourth Generation Evaluation* argues for the importance of human, political, social and cultural context of those taking part in the evaluation. Responsible technology professionals, therefore, will be aware of possible conflicts of interest and try to avoid them. Increased attention to participatory approaches to technology assessment to accommodate a broader range of considerations when assessing new technologies. For this reason, a technoethical framework based on conversation theory (CT) is selected for its emphasis on multiple-perspective sharing and consensual decision making. This choice is validated by previous applications of conversation theory in complex technology assessment contexts (Laurillard, 1999) .

Conversation Theory (CT) was developed by Gordan Pask in the 1960s and 1970s to explain the conditions needed for human and artificial systems to learn and create shared knowledge by modeling learning conversations (Pask, 1961; 1975). Boyd (2004) states," Conversation theory portrays and explains the emergence of knowledge by means of multilevel agreement-oriented conversations among participants, supported by modeling facilities and suitable communication and action interfaces" (p.179). A technoethical framework based on CT specifies the minimal

Figure 1. Learning conversation model for technology assessment

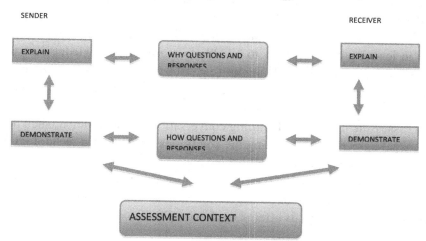

conditions required for assessing technologies in education and work. It does so by applying a learning conversation model to promote mutual understanding and consensual decision-making (Luppicini, 2008). Modeling learning conversations in technology assessment helps to prevent technology assessments from being biased towards one particular perspective. It also helps guard against possible ethical and legal battles that arise due to stakeholder disputes. Learning conversations are situated within the conversational network of the stakeholder group and thus, provide a means of establishing how (demonstrate) and why (explain) ethical considerations are relevant within a specific context of stakeholder interest.

As depicted in Figure 1, a learning conversation model illustrates the basic conditions needed by each stakeholder to create a collective understanding of all stakeholder perspectives required for informed consensual decision-making with regard to technology. This requires that (1) a receiver shares a concept with the sender when the receiver is able to make sense of the sender's explanation (demonstrate), and (2) the receiver and the sender understand one another when the receiver is able to make sense of the sender's explanations of the receiver's explanations (explain). In the most basic learning conversation model there is a minimum of two levels of interaction to leverage mutual understanding/knowledge of the how and why of a topic (learning conversation).

Learning conversation models are highly suitable in complex contexts where technology has an influence or mediating role due to the complexity of the situation, presence of multiple stakeholders, and a high-level of information sharing. Conversation theory treats human and machine systems as self-producing agents creating and re-creating knowledge by continually telling themselves what is known (self-reproduction). The process of self-reproduction explains how knowledge is created (sender) and reproduced by another (receiver) by establishing joint conversations of mutual agreement over what is known. In this case, the stakeholders must engage in learning conversations to be able to reproduce knowledge sent from a variety of technology mediated sources (e.g., documents, artifacts, diagrams, statistical tools etc.) to create knowledge shared with other stakeholders in an effort to reach consensus based on multiple-perspective sharing embedded in a shared context (assessment context). Learning conversations are completed when there is a consensual agreement concerning the conditions needed for a technology to be acceptable to all stakeholders. Establishing consensual agreement on controversial areas of technology assessment (i.e., stem cell research, nuclear technologies, cloning) can draw on a variety of strategies to evaluate stakeholder arguments. One particularly useful strategy is offered by Tavani (2007):

- **Step 1:** Convert the argument into standard form (List all premises, followed by the conclusion).
- **Step 2:** Test the argument for its strength of reasoning to see whether it is valid or invalid.
- **Step 3:** Is the argument valid? If yes, go to step 4. If no, go to step 5.
- **Step 4:** Is the (valid) argument also sound? That is, are the premises true in the actual world?
- **Step 5:** Is the (invalid) argument inductive or fallacious?
- **Step 6:** Determine whether the premises in your argument are either true or false in the actual world.
- **Step 7:** Make an overall assessment of the argument by determining both (a) the argument's strength or reasoning (valid, inductive, or fallacious) and (b) the truth conditions of each or the argument's premises. (p. 85-86)

Given the controversial nature of many areas of technology innovation and the presence of political influences, participatory approaches with a high level of stakeholder interaction and an eye towards ethical and social considerations are highlighted as core to effective technology assessment.

DISCUSSION QUESTIONS

- What are the consequences of electronic technologies on core principles and values that guide our face-to-face social relationships?
- How can educators teach that it is unacceptable to download something that

is copyright protected and claim it as one's own creation?

- What ethical responsibilities do educational institutions have to ensure that all students have equitable access to educational technology?
- What should schools do to ensure appropriate Internet use in the classroom?
- What is an appropriate ethical framework within which to educate students within a technological environment?
- Do educators have a moral and professional duty to submit students' work to an Internet plagiarism detection services? Should students be able to refuse to have their work submitted to Internet plagiarism detection services?
- What responsibilities to educators have to teach about the ethical use of educational technology? Do students understand what ethical behavior is, especially ethical behavior when using technology?
- Whose responsibility is it to ensure that educational institutions adequately preparing students to use technology ethically at school and work?
- Researchers and technologists often lack formal training in ethics and technology. How does this influence the development of science and technology?
- Elected officials often lack formal training in ethics and technology. How does this influence the development of science and technology?

DISCUSSION

Educational technoethics, as it is presented in this chapter is treated as an important branch of technoethics dedicated to social and ethical aspects of technology within traditional educational settings as well as professional educational contexts. General educational technoethics is concerned with ethical use of technology to promote the aims of education while professional technoethics is a specialized area of educational technoethics focusing on the development and evaluation of ethical codes and standards to guide decision-making about technology in education and professional life. The integration of new technologies in education has given rise to new ethical dilemmas concerning the digital divide, information privacy, censorship, cyberbullying, and plagiarism. Within professional training, the need to go beyond technical expertise to deal with ethical considerations has been firmly established and the need for professional technoethics substantiated. There is now a recognized need for ethical considerations within the core training for technology professionals and the creation of technological assessment tools applied in society. Skorupinski (2002) argued, "Scientific expertise and moral competence are not the same thing and must not be equated with each other. One aim of discourse on the acceptability of risk should be the development of mutual understanding regarding the perception and evaluation of risk" (102). The author highlights the context-sensitive nature of technology in society and the need to communicate values, norms, and evaluation criteria to the public. This, in turn, can help establish a mutual understanding needed for participatory decision-making about new technology. CT and conversation modeling strategies highlight the important role of principles and practices which allow for ethical considerations to be integrated through stakeholder participation and communication processes with the goal of achieved mutual consensus in assessing new technologies.

Based on work covered in this chapter, there is a demonstrated need for continued work in educational technoethics to address key issues arising from the ongoing challenges of integrating technology into education and training. This book follows Cortez (2008) recommendation that "Without a technoethical framework, we are then left with a system of education and a society immersed in efficiency, technological growth, a multitude of technology artifacts, and information saturation"(664).

REFERENCES

Ashworth, P., Bannister, P., & Thorne, P. (1997). Guilty in whose eyes? University students' perceptions of teaching and plagiarism in academic work and assessment. *Studies in Higher Education, 22*(2), 137–148. doi:10.1080/03075079712331381034

Boyd, G. (1991). The shaping of educational technology by cultural politics, and vice versa. *Educational and Training Technology International, 28*(3), 87–96.

Boyd, G. (2004). Conversation theory. In D. Jonassen (Ed.), *Handbook of educational communications and technology (pp.178-198)*. Mahwah, NJ: Laurence Erlbaum and Associates.

Candor, J. (2008). The ethical dilemma over money in special education. In R. Luppicini & R. Adell (eds.), *Handbook of research on technoethics,* (pp. 409-425). Hershey, PA: Idea Group Publishing.

Cortez, J. (2008). Historical perspective of technoethics in education. In R. Luppicini & R. Adell (eds.), *Handbook of research on technoethics,* (pp.651-667). Hershey, PA: Idea Group Publishing.

D'Abate, C. P. (2005). Working hard or hardly working: A case study of individuals engaging in personal business on the job. *Human Relations, 58*, 1009–1032. doi:10.1177/0018726705058501

Donath, J. A. (1999). Identity and deception in the virtual community. In M. A. Smith & P. Kollock (Eds.), *Communities in cyberspace,* (pp. 29-59). London: Routledge.

European Parliamentary Technology Assessment. (2008). Retrieved June 15, 2008, from http://www.eptanetwork.org/EPTA/about.php

Gearhart, D. (2008). Technoethics in education for the twenty-first century. In R. Luppicini & R. Adell, (Eds.), *Handbook of research on technoethics,* (pp.263-277). Hershey, PA: Idea Group Publishing.

Guba, E., & Lincoln, Y. (1989). *Fourth Generation Evaluation.* San Francisco: Sage Publications.

Gurak, L. J. (2001). *Cyberliteracy: Navigating the Internet with awareness.* New Haven, CT: Yale University Press.

Haghi, A., Mottaghitalab, V., & Akbari, M. (2008). The scholarship of teaching engineering. In R. Luppicini & R. Adell (eds.), *Handbook of research on technoethics* (pp. 439-452). Hershey, PA: Idea Group Publishing.

Lathrop, A., & Foss, K. (2000). *Student cheating and plagiarism in the Internet era: A wake-up call.* Englewood Cliffs, NJ: Libraries Unlimited.

Laurillard, D. (1999). A conversational framework for individual learning applied to the 'learning organization' and 'learning society.' . *Systems Research and Behavioral Science Systems Research, 16,* 113–122. doi:10.1002/(SICI)1099-1743(199903/04)16:2<113::AID-SRES279>3.0.CO;2-C

Lim, V. K. (2002). The IT way of loafing on the job: Cyberloafing, neutralizing and distraction versus destruction. *Cyberpsychology & Behavior, 9*(6), 730–741.

Luppicini, R. (2005). A systems definition of educational technology in society. [Cambridge, MA: MIT Media Lab.]. *Educational Technology & Society, 8*(3), 103–109.

Luppicini, R. (2008a). The emerging field of technoethics. In R. Luppicini & R. Adell (eds.), *Handbook of research on technoethics (pp. 1-18).* Hershey, PA: Idea Group Publishing.

Luppicini, R. (2008b). Introducing conversation design. In R. Luppicini (Ed.), *Handbook of conversation design for instruction (pp.1-15).* Hershey, PA: Idea Group Publishing.

Muffoletto, R. (2003, November/December). Ethic: A discourse of power. *Tech-Trends, 47*(6), 62–67. doi:10.1007/BF02763286

Nichols, R. G. (1987). Toward a conscience: Negative aspect of educational technology. *Journal of Visual/Verbal Languaging, 7*(1), 121-137.

Pask, G. (1975). *Conversation cognition and learning: A cybernetic theory and methodology.* Amsterdam: Elsevier.

Publications, O. T. A. (2008). Retrieved February 17, 2009, from http://www.princeton.edu/~ota/ns20/pubs_f.html

Pullen, D. (2008). Technoethics in schools. In R. Luppicini & R. Adell (eds.) (2008), *Handbook of research on technoethics,* (pp.680-698). Hershey, PA: Idea Group Publishing.

Redding, W. C. (1996). Ethics and the study of organizational communication: When will we wake up? In J. Pritchard (Ed.), *Responsible communication: Ethical issues in business, industry, and the professions,* (pp. 17-40). Cresskill, NJ: Hampton Press.

Robbins, R., Fleischman, K., & Wallace, W. (2008). Computing and Information Ethics Education Research. In R. Luppicini & R. Adell (eds.), *Handbook of research on technoethics,* (pp.391-408). Hershey, PA: Idea Group Publishing.

Ryder, M. (2008). The cyborg and the noble savage: Ethics in the war on information poverty. In R In R. Luppicini & R. Adell (eds.), *Handbook of research on technoethics (pp.232-249).* Hershey: Idea Group Publishing.

Socol, I. (May, 2006). Stop chasing high-tech cheaters. *Inside HigherEd.* Retrieved February 17, 2009, from http://www.insidehighered.com/views/2006/05/25/socol

Tavani, H. T. (2007). *Ethics and technology: Ethical issues in an age of information and communication technology* (2nd Ed.). Hoboken, NJ: John Wiley & Sons.

Turnitin. (2007). *Turn it in.* Retrieved January 12, 2007 from http://www.turnitin.com/static/home.html

Underwood, J., & Szabo, A. (2003). Cybercheats: is information and communication technology fuelling academic dishonesty? *Active Learning in Higher Education, 5,* 180–199.

Wilson, C. (2005). *Computer attack and cyberterrorism: Vulnerabilities and policy issues for Congress.* Retrieved June 1, 2009, from http://www.ipmall.piercelaw.edu/hosted_resources/crs/RL32114_050401.pdf

ADDITIONAL READING

Alge, B. J. (2001). Effects of computer surveillance on perceptions of privacy and procedural justice. *The Journal of Applied Psychology, 86*(4), 797–804. doi:10.1037/0021-9010.86.4.797

Griffiths, M. (2003). Internet abuse in the workplace: Issues and concerns for employers and employment counselors. *Journal of Employment Counseling, 40*(2), 87–96.

Marcuse, H. (1964). *One dimensional man.* Boston: Beacon Press

Marshall, K. (1999). Has technology introduced new ethical problems? *Journal of Business Ethics, 19*(1), 81–90. doi:10.1023/A:1006154023743

Orwell, G. (1949). *1984.* Harcourt Brace Jovanovich, Inc.

Reich, C. (1970). *The greening of America.* New York: Random House

Section 4
Globalization and Technoethics

Chapter 10
Global Technoethics and Society

INTRODUCTION

As discussed in Chapter Five, Technoethics and Society is a branch of Technoethics concerned with the ethical use of technology to promote the aims of contemporary society. This includes the study of local organizations (micro level) as well as the study of broader global structures and processes (macro level). Although this section focuses mainly on global technoethics (macros level), it provides a sketch of organizational and global developments to better situate the discussion. Since the beginning of the 20th century, work in organizational studies reflects a continually evolving research area within communications and other fields (human resource management, industrial relations, business management) since the contribution of early classical models of organizations. Classical models provided a new technology

DOI: 10.4018/978-1-60566-952-6.ch010

(technique) for organizational leaders to help manage increasing large contingencies of workers within a top-down communication structure geared towards maximizing organizational performance and efficiency. The coming of the information age and the development of sophiosicated information and communication technologies (ICT's) provides additional technology for transforming society and the institutions within it. This was the start of the trend towards the modern day multi-corporation. Continuing progress through human relations and human resource approaches, systems theory, and ideological perspectives (critical, cultural, and feminist) on organizations have reinforced this core work. Rapid technological developments within organizational life, the increasing power of multi-corporations, and the faster pace of organizational change processes, have helped nurture in globalization

BACKGROUND: FROM ORGANIZATIONAL TO GLOBAL TECHNOETHICS

Technology and Organizational Transformations

Organizational technoethics focuses on how technological advances are redefining organisations and how they operate within an evolving knowledge economy. The shift of organizations to greater knowledge creation combined with the integration of technological advances has led to changing working relations. Beginning in the 1980s, advances in ICT's, have transformed the nature of information sharing within and between organizations while offering new opportunities and increased efficiency. This is particularly true in areas of the new knowledge economy dependent on ongoing technological innovation and scientific knowledge. The ability to classify and exchange multiple types of data (text, visual, audio) is now possible via the Internet. Because of the copy and paste capacity of Internet data exchange, it is relatively simple to create endless combinations of new data from existing data. The increased malleability of information within and ICT environment has leveraged knowledge creation and information exchange in many work teams dependent on knowledge creation (Wouters & Schrodder, 2003). Wouters and Schrodder (2003) provide up to date coverage of key issues and challenges associated with the use of ICT for data sharing. They draw attention to the new possibilities created by the institutionalized use of ICT for scientific data sharing:

Science once was about scientists creating knowledge by researching nature. As time passed science grew in size and complexity and became a more co-operative and cumulative affair. Science developed a Global Science System involving larger teams of scientists (human resources) using larger instrumental facilities (capital goods) to study aspects of nature (data) in order to disseminate knowledge (product) (p.7).

It is this multi-actor global scope of data sharing that differentiates data sharing of today from other periods in history in terms of breadth and depth of data sharing potential. At the same time, this expanded breadth and depth create additional challenges:

Instead of considering data sharing as getting free additional help in getting the intended scientific work done, initial investors in digital data are often suspicious of unfair competition and free riding. Researchers looking for existing data sources from colleagues are not always welcomed heartily. Researchers, officials and managers mention financial, legal, and organizational barriers, cultural and ethical problems that complicate the full realization the potential of digital data resources (p. 9).

New advances in instrumentation and networking have also transformed research practices and the nature of knowledge creation within the research and development of many organizations. In particular, this has leveraged research work on and with the Internet where Internet provides its own observable and controllable (to some extent) environment. Schilling (2000) noted how the combination of research instrumentation with Internet-based communication tools in scientific research is altering research and how it is conducted. The author likened Internet environments to a virtual observatory with a high level of opportunity for observation and measurement. In terms of research processes, ICT's have changed how many researchers work together. The increasing centrality of ICT is closely linked to contemporary academia as a key tool in academic work for knowledge creation. This is particularly salient in the domain of science and technology research and development where large scale projects involving distributed work teams are transforming the very nature of research culture and strategies. According to Merz (2006):

In the future, e-Science will refer to the large scale science that will increasingly be carried out through distributed global collaborations enabled by the Internet. Typically, a feature of such collaborative scientific enterprises is that they will require access to very large data collections, very large scale computing resources and high performance visualisation back to the individual user scientists (p. 100).

Organizational advances via technology enabled advances in information sharing and distributed scientific collaborations is a small part of the growing knowledge economy which is redefining organisations and how they operate. Knowledge creation has become entrenched within organisations attempting to stay competitive. Haag, Cummings, McCubbrey, Pinsonneault, and Donovan (2006) estimated that the growing knowledge economy represented 75% of total work within North America. On the human side, the shift of organizations to greater knowledge creation combined with the integration of ICT's also allows more flexible work relations in organizational operations. This has led to changing working relations through telework (working from home), videoconferencing, shareware applications within traditional organizations. This has also led to the establishment of virtual organizations and virtual organizational alliances with all organizational operations conducted at a distance. As is discussed below, this expansion of the organizational landscape did not come without a price.

Corporate Crime is perhaps the most challenging problem within contemporary organizations and society. This is largely due to the added flexibility and opportunities offered by new technologies within organizations and society. Technological advances in the 1980s and 1980s provided the context for an emerging knowledge based economy dominated by a growing number of multi-corporations able to capitalize on the cost reduction and increased efficiency offered by an increased economy of scale and the availability of new ICT's and transportation technologies. However, in accordance with the Law of Technoethics, social and ethical issues also expanded with major legal and ethical battles arising within organizations all over the world including: HP (Spying Scandal), Bre-X scandal (stock scandal), Clearstream (money laundering), Enron (accounting fraud), Firestone Tire and Rubber Company (child labor violations), Guinness (share trading fraud), Halliburton (war profiteering), Nortel (criminal fraud), Tyco International (embezzlement), Worldcom (fraud) and other companies around the world (Austen, 2008; Corpwatch, 2003; Goold & Willis, 1997; Jeter, 2003; Kaplan, 2006; Kochan & Pym, 1987; MacLean & Elkind, 2002; Robert & Backes 2002; Salinger, 2004; United Nations Mission in Liberia, 2007; USA Today, 2005; York, 2003). It is not only corporations which have been the focus of ethical scrutiny. Steve May's (2006) edited *Case Studies in Organizational Communication: Ethical Perspectives and Practices* provides a collection of ethical case studies on a variety of issues within the military, universities, religious institutions, multi-corporations, small businesses, public services, and hospital settings. For a more comprehensive view of the current state of organizational abuses and criminal activity, see Salinger's (2004) *Encyclopedia of White-collar & Corporate Crime*, which covers even broader coverage of the plethora of legal and ethical issues which have led to a schism between many organizations and the public.

Technology and Organizational Cultures

The technology driven trend in organizational transformation can be juxtaposed with other efforts in the 1980s and 1990s to re-invent organizations as communities of practice closely connected to organizational learning, team building, and knowledge management. How does the contemporary organizational culture of technology affect organizational values, knowledge creation, and communications? What are the social and ethical considerations within a community of practice for knowledge creation? A number of studies have examined science and technology as social practice and subject to the culture present within professional work (Knorr-Cetina, 1999; Latour & Woolgar, 1986). Within Science and Technology Studies (STS), Latour and Woolgar (1986) explore the culture of scientific work in how scientists shape the co-construction of facts and meaning:

If facts are constructed through operations designed to effect the dropping of modalities which qualify a given statement, and, more importantly, if reality is the consequence rather than the cause of this construction, this means that a scientist's activity is directed, not toward "reality," but toward these operations on statements (p. 237).

In a similar vein, Knorr-Cetina's (1999) *Epistemic Cultures: How the Sciences make Knowledge* offers an ethnographic exploration of how epistemic culture shape work practices, institutional organization, and power relations through the use of technologies. This work helps reveal how there is a culture (or cultures) of technology at work in the study of contemporary organizations that brings with it a social and ethical character with social practices and rituals particular, what Wenger (1998) labeled 'communities of practice.' Communities of practice in science and technology constitute a major area of knowledge production within the evolving knowledge economy that constitutes to grow. Communities of practice describe learning processes and socio-cultural practices arising through the co-construction of knowledge (and meaning) among a group of individuals working together with common goals (Wenger, 1998). Within various organization oriented literatures (organizational development, organizational communications, knowledge management) communities of practice for knowledge are recognized as a means of leveraging social capital, encouraging knowledge creation, and facilitating the recovery of tacit knowledge within an organization (Dalkir, 2005).

In attempting to sketch out the notion of 'cyberscience' Nentwich's (2003) *Cyberscience: Research in the age of the Internet* examines how the concentrated use of ICTs in technology oriented fields is continuing expanding and transforming publishing practices, research collaboration tendencies, and communication sharing. This work was based on extensive interviews with experts in the field. This work helps to draw out the importance of social and cultural factors influencing actual organizational work in science and technology. The cultural aspects of organizational life raises social and ethical issues connected to the nature of science and technology within the knowledge economy. What are the ethical considerations in revealing scientific finding which may negatively influence the community of practice? What social and ethical responsibilities do scientists and technologist have to inform the public of about dangers connected to their work? What are the social and ethical values within communities of practice for knowledge creation within controversial areas of science and technology?

This cultural convergence brings with it a number of social and ethical considerations to deal with. This is partly due to the complexity of technology and challenges tracing its multifaceted influence on individuals, organizations, nations, and global audiences and other stakeholders. This is also partly due to the lack of communication, miscommunication, and control of communication. This work raises important questions concerning truth and ethics which organizations must deal with. To what extent is science manipulated to benefit organizational stakeholders? How is access to various epistemic cultures (science, technology, research and development) managed, by whom, and for what purposes? Why are people in need of existing interventions (e.g. AIDS treatments) often cut off from this, along with the knowledge and expertise required to create their own interventions?

The Background of Global Tech

Globalization, as it has evolved during the last fifty years, is a controversial area of research and development marked by rapid technological change, heated debate, and a lack of understanding about its potential and actual contribution to society. Although, scholarly attention to globalization can be traced to the works of Karl Marx, the term 'globalization' did not come into popular use until the 1960s and 1970s when political and economic interdependence between states led to closer interconnections between separate states and people (Modelski, 1972). This interconnectedness was facilitated by rapid technological innovation that transformed the nature of economic growth, international relations, community involvement, communication, social interaction, language, and culture. In the 1960s and 1970s the economic drive for globalization was nurtured in by the advent of global telecommunications and transport technologies along with technology driven industries (i.e., biotechnology, computing, genetically modified foods). At the same time, technology was met by opposition. This was partly due to the negative consequences associated with technological growth and partly due the exaggerated benefits of technology that were not realized. By the 1990s, the amassing body of scholarly work and debate made it one of the most significant interdisciplinary and multidisciplinary areas of study intersecting the fields of sociology and anthropology, economics, international relations, communications and cultural studies, business studies, and technology studies.

Technological Change and the Global Economy

Up until the 1950s, economists equated land, labor, and capital as the chief components and measures of economic productivity and growth. However, the development of telecommunications and ICT's nurtured in the information revolution and the creation of a technology driven knowledge economy (often called intangible or residual factors) not considered in traditional economic measures. This shift to a knowledge-based economy is linked to both advances in new material technologies (telecommunication systems) and process technologies (organizational principles and skills development) helped nurture in a global economy.

The technology driven trend towards a global economy has created a number of subversive consequences within the economic realm and beyond. First, the development of new technological products typically requires a fairly large infrastructure with adequate resources to support its production. This has created vulnerable single industry towns around North America, entirely dependent on the sustainability of this industry for economic prosperity. The downsizing of multi-national corporations during the 2008 economic recession like General motors and Hershey have placed huge strains on local Ontario communities like, Oshawa and Smith Falls, not to mention surrounding communities with connections to these industries. Volti (2006) explains, "Many places of employment have closed down as new products and processes have replaced old ones, leaving communities and their inhabitants in desperate straits" (p.20). Second, the rapid pace at which new products supplant old products in the drive for technological progress, can lead to the loss of culturally entrenched human activities and practices (e.g., the art of letter writing, traditional journalism, railway travel, etc.). This subversive aspect of technology spills over into other areas of life and society. Volti (2006) notes that "technological change is often a subversive process that results in the modification or destruction of established social roles, relationships, and values (p.18). This demonstrates an important connection between the global economy and its social life difficult to ignore.

Information Technology and New Development Opportunities

The information revolution and the establishment of a global telecommunications infrastructure that transformed information and communication exchange around the world. The rapid spread of the information revolution transformed the nature individuals and societies interacted, exchanged information, and communication. Advancing ICT's led to space-time compression by allowing more flexible exchanges across vast geographical distances. These developments transformed the landscape of social interaction and the economy around the world by quickening the pace of interactions, accelerating the interconnection of separate economies and societies, while eroding geographic borders that constrained social interactions. The developments driven of information technology advances were especially important for propelling the global economy as indicated by the 1996 *World Investment Report*:

Rapid technological developments in telecommunications and computers in the 1980s have made some services, especially information-intensive ones, more tradable by providing 'the means for overcoming the inherent obstacle to trade in many services ... the intangibility, non-storability, and hence non-transportability of these services UNCTAD (1996, p. 105).

James' (2006) *Globalization, Information Technology and Development* provides an insightful exploration of the close connection between globalization and information technology. It builds on investigations carried out on developing countries arguing that information technologies are associated with a number of powerful cumulative mechanisms causing some countries to grow rapidly and others to become increasingly marginalized from the global economy. James explores how information technology policies promote globalization by: (1) reducing communication costs and information imperfections, (2) enhancing comparative advantage of technology adopting firms and countries at the expense of non-adopters, and (3) increasing trade export of electronic devices (e.g., computers, PDA's etc.), and (4) promoting partnerships and alliances between multinational corporations. In this sense, the information revolution has molded modern globalization as primarily a technological phenomenon driven by technological innovation with resulting influences placed on international trade and foreign investment within an emerging global economy.

Scientific Knowledge and Technological Advances

A big part of the emerging global economy was connected to advances in science and the use of scientific knowledge and expertise to fuel technological advances. This, in turn lead to increasingly close connections between science and technology unlike previous periods in history. Although traditionally, science and technology have complimentary roles, the relationship between scientific knowledge and technological innovation is complex and changing. More specifically science and technology have converged more and more in terms of their mission and mandate than ever before. Traditionally, science and technology were treated as separate enterprises with different natures and goals. Volti (2006) explains, "Whereas science is directed at the discovery of knowledge for its own sake, technology develops and employs knowledge in order to get something done" (p.57). In other words, science focuses on finding the truth (Is it true or not?) and technology focuses on providing practical solutions to posed problems (Does it work?). Within the global marketplace, however, there has been a shift in science from a pure to a more practical orientation similar to that which drives technological innovation. This shift is due to a complex variety of factors which provide science its legitimacy. As noted by Volti (2006):

Financial support, however important it is to the maintenance of scientific inquiry, is only part of the picture. The willingness of government agencies to grant money for scientific research and of citizens to have their taxes used in this manner is indicative of a widespread belief in the in the legitimacy of scientific research. This legitimacy is in large measure the product of the presumed ability of science to ultimately produce practical results (p. 61).

The Propaganda of Global-Tech and Its Misgivings

Although technology (and science) may provide solutions to challenges in work and life, there are many areas where the promises of technology have fallen short. This is partly due to miscommunication from certain groups who have greatly exaggerated the potential benefits of technology and underestimated its limitations. There has been a longstanding appeal to the power of technology and multiple appeals to technological innovation and technocratic approaches for transforming many aspects of work and life. This was influenced by popular technocratic views on society and governance, along with more focused programs of organizational management, the appeal to theories of Scientific Management spearheaded by Frederick Taylor in the early 20th century. These programs emphasized the management of organizations and society on technical and scientific principles created by experts to create policies for governing. But, as noted by Volti (2006), technocratic based systems like Scientific Management, often overlook the differences between and socio political problems, "The basic fallacy of Scientific Management, one shared by all other variants of technocracy, is that administration can replace politics" (p. 30).

Another factor that contributed to the proliferation of propaganda was the widely held assumption that technological advancement and progress in society were synonymous. However, it is not the case that technological advances always lead to progress or that the newer technologies always the best. There are a number of serious flaws in how technology has been presented as a solution for almost everything. There are number of problems with this to note. First, there is the problem of illusion. The illusion of technology is that can think when, in fact, it cannot. Although technologies can be strategically applied to produce a desired effect, which does not mean that the mechanisms through technology operate are understood. The tendency to exaggerate the power of technology can have dire consequences. For instance, in the case of employing certain drugs to alleviate depression may help block out thought patterns and emotions which prevent functioning in society. This does not mean that it eliminates the problem. As noted by Volti (2006), "Above all, technological solutions only eliminate the surface manifestations of the problem and do not get at its root (p. 24). Certainly, a pharmaceutical fix will not replace years of therapy needed to get at the real root of someone's depression. The second problem is the one-size fits all approach to technology application. Many technologies are effective in some contexts but not others. The reason is that technologies are emerged in society and do not exist independent of the individuals who make and utilize them. Third, technology can also create more serious problems in other areas of life and society in solving its problem. Consider the ethical controversies, and public protests that have taken place concerning life-preserving technologies and enhancing technologies. It is possible to keep people alive with the aid of technology (e.g., dialysis machine) when they could not otherwise survive. It is possible to screen out hereditary conditions to ensure healthier babies. These are controversial areas that would be best addressed before a technology is publically applied. The way to avoid falling into any of these problems is to appreciate the relational system of technology which is interwoven in human activity and society.

The Global-Tech Backlash

The integration of technology into modern work and life is not without its critics. It must be juxtaposed with efforts to push it back. The early Luddites movement in 19th century England was led by Ned Ludlum in opposition to technological advances perceived to eliminate jobs. The intentional sabotage of machines by workers was a social movement that spread as a form of worker protest. Similar protest movements from workers in the late 1970s (labeled neo-luddites) were repeated in the newspaper industry in the mid-1970s by linotype operators in the United States (Volti, 2006). As computer technology evolved during the 1980s and 1990s, more subtle opposition to technology (and how it is used) came from hackers with technology expertise who used technology against technology, often as a weapon to negatively affect corporate control. Industrial sabotage from members of environmental movements has also attempted to push back certain uses of technology. For instance, in the 1990s, tree spiking by environmental zealots was done to oppose clear cutting of B.C. forests. This injured workers in the lumber industry. Recent pipeline bombing of sour gas pipelines in Dawson B.C. to protest the sour gas industry has resulted in industry costs and setbacks.

Technological driven threats (i.e., mass terrorism, nuclear weapons, outsourcing and worker exploitation) have also cropped up with within As stated by Held and McGrew (2003), "Transnational and transborder problems such as the spread of genetically modified foodstuffs, mass terrorism and money laundering, have become increasingly salient, calling into question the traditional role, functions and institutions of accountability of national government "(p. 38). Technological driven threats have extended the range of criminal activity around the world. Findlay's (2000) *The Globalization of Crime: Understanding Transnational Relationships in Context. Cambridge* examines the dark side of globalization and the emergence of crime as intertwined in global trends. This text attempts to examine the role of criminal activity in a global context and present an integrated theory of crime and crime and social development, social control and the political economy of crime in order to understand the role of crime in social change. Findlay (2000) states, "In a globalised world there will be a single society and culture occupying the planet. This society and culture will probably not be harmoniously integrated although it might conceivably be. Rather it will probably tend towards high levels of differentiation, multicentricity and chaos" (p.2). This work provides convincing evidence of crime as a major force in globalization, highlighting the disproportionately high level of criminal activity among individuals who "have experienced high residential mobility" p.27). This dark side of globalization and its long-term consequences is a very real part of globalization that cannot be ignored.

Illusions of Global Measurement

Based on the above review of technology and globalization, it is apparent that there is a disconnect in how globalization has evolved. One of the greatest challenges in understanding globalization lies in a schism that exists between how globalization is defined and how it is measured. Although the predominant conceptions of globalization view t as a developing process with many dimensions (i.e., economic, educational, political, cultural, communications, education), there is a paucity of work on non-economic dimensions of globalization. This is partly due to the longstanding tradition of macro level economic analyses of globalization since the 1960s, over 20 years before globalization became a major area of multidisciplinary study. It is also due to the more easily available sources of economic data compared to data on other dimensions of globalization. Many data sources on non-economic aspects of globalization are not calculated and other more detailed studies of globalization which highlight relational aspects of globalization are not available. For instance, measuring global and national trends in Internet saturation does not get at the local differences in culture, politics, education, and economy that will have a mediating influence. This would require more in-depth naturalistic studies of globalization which are more costly and time consuming, This has given a skewed perception of globalization and ignored important interactions of globalization having a major impact in the world. This can be likened to losing one's keys at night in the bushes but only looking around the street light where the area is well lit. Globalization has many sets of lost keys in the bushes waiting to be discovered. Approaching globalization as

THEORIES OF GLOBALIZATION

Globalization Theories

World-systems theory, World culture theory, and world polity theory are general globalization theories cited the most in current globalization literature. World-systems theory defines globalization as a historical process started in the 15th century that gradually allowed the capitalist world-system to spread globally. According to Wallerstein (1998), globalization is not new, but rather, views the "ideological celebration of so-called globalization is in reality the swan song of our historical system" (p. 32). One of the main strengths of globalization research under this theory lies in the breadth of its application and easy to operationalize units of analysis applied to any bounded historical social system of interdependent parts that operate according to distinct rules. Within the current context, world-systems theory views globalization in terms of a single capitalist world-economy with no political center because it " has had within its bounds not one but a multiplicity of political systems." This has provided capitalism the "freedom of maneuver that is structurally based" (p. 390). Major limitations of this theory concern the lack of specificity in defining the economic parts (economy seen as single unit with a division of labor) along with a lack of attention to other important dimensions of globalization (politics, culture, communication).

World culture theory, spearheaded by Robertson (1992), views globalization as the realization of world as a whole comprised of four main elements and corresponding processes: (1) nation-states and societalization processes, (2) the system of societies and internationalization processes, (3) individuals and individuation processes, and (4) humankind and the generalization of consciousness about humankind. Unlike the World-Systems Theory, which examines historical processes, the world-culture theory attempts to trace how the world has gravitated towards a global city (unicity). One major strength of this theory is that it provides much needed focus on the cultural dimension of globalization by exploring how multiple cultures struggles to coexist in a single time and place. Major limitations include a lack of consideration of other dimensions of globalization and a lack of focus on specific structures which influence globalization. Some limitations are overcome by the world polity theory which also conceptualizes globalization in terms of the growth and activity of world culture. However, world polity offers a more pragmatic focus on modern globalization as constituted by a rationalized institutional and cultural order that follows universally applicable models to influence nations, organizations, and individual identities (Meyer et al., 1997) . Conceptions of progress, sovereignty, rights, and the like, have acquired great authority, structure the actions of states and individuals, and provide a common framework for global disputes. According to Meyer et al. (1997) there are four main elements of collective world society that help shape world culture: (1) international governmental organizations, (2) nation-states, (3) voluntary associations, and (4) scientists and other professionals dealing with world-cultural issue. Nation-state identities and structures are influenced by collective world society values. Meyer (1997) states, "As creatures of exogenous world culture, states are ritualized actors marked by intensive decoupling and a good deal more structuration than would occur if they were responsive only to local, cultural, functional, or power processes" (p,. 173). The major drawback of world polity theory lies in its assumption that world culture is highly rationalized with rational actors when this is rarely the case. Taken together, general globalization theories conceptualize globalization in narrow terms which fail to get at its multidimensional nature.

Technology Oriented Theories

Technology oriented theories of globalization also contribute to our understanding of globalization through its connection to technology assumed to be a major driver of global change. A number of useful theories exist which help to explain various facets of technology and globalization including globalization theories, technological determinism, social constructivism, actor-network theory, diffusion of innovation,

Technological determinism focuses on how material and physical laws cause technology to develop in certain ways independent of human control. It assumes that technology shapes human activity in society and limits free will. Ellul's(1964) *The Technological Society* is a seminal work based on *technological determinist assumptions*. Ellul's work warns of the dangers in modern technological society to enslave people by systematically impose technology at the expense of human autonomy. He stated, "Technique enslaves people, while proffering them the mere illusion of freedom, all the while tyrannically conforming them to the demands of the technological society with its complex of artificial operational objectives" (35). The major strength behind its approach is that it helps address the impact of technological development on human activity and society. The major limitation with this approach is that it is one sided, exploring only technology while leaving other important factors (human, social, political, economic) unaddressed.

Various social constructivists focuses on the importance of social groups and flexibility in technological change. **Social construction of technology** (a **SCOT**) is a variation of social constructivist theory which emphasizes how human action shapes technology embedded within a social context. For example, Pinch and Bijker (1987) applied SCOT to explain how stakeholder groups helped shape the development of a new technology (bicycle design) that included technological design elements and other non-technical influences such as dress code considerations, public appeal, and safety issues. Another leading approach, Structuration theory is a social constructivist oriented perspective developed through the work of Anthony Giddens (1986). It focuses on systems (outcomes of interactions between individuals) and structures used by individuals to negotiate social life including: signification (rules of language and communication), domination (power relations connected to the distribution of resources), and legitimation (moral rules and behavioral norms). Structuration describes the ongoing interaction between system and structure. For instance, Desanctis and Poole (1994) applied structuration theory to analyze the complexity of structured relied on when using advanced technology within a social system. In the case of technology and globalization, structuration theory might deal with how people follow structures that influence human practices with technology applied to some area of globalization. The major advantage of these social constructivist theories lie in their ability to address individual and social influences which shape technology. The major limitation is that social constructivist theories is not so much a theory as a set of general principles that guide almost all forms of social inquiry. A recent social constructivist derivation called actor-network theory (ANT) has provided attempted to provide more detailed explanations of how social construction takes place.

Actor-Network theory (ANT) examines social situations in terms of the various types of relationships emerging in connection to technology. Actor-network theory (ANT) draws on insights from qualitative methodology (ethnomethodology, grounded theory, case study) and STS scholarship focused on the study of innovation processes and knowledge-creation within technological systems. ANT was advanced in Science and Technology Studies (STS) by Michel Callon and Bruno Latour as a 'material-semiotic' method intended to trace material relations (between things) and semiotic relations (between concepts). ANT assumes that material and semiotic relations form a single network and that networks are often transient, with changing relations, and conflicts linked to social, political, organizational, legal, technical and scientific factors. From the 1990s onward, ANT applications spread beyond STS scholarship and is now applied as a tool for analysis in a organizational communications, informatics, health studies, sociology, and economics.

In terms of methodological application, the general aim of ANT is to explore how material and semiotic relations comes together within networks, how networks sustain themselves, and how they degenerate. Key concepts in ANT include the notion of translation and heterogeneous network. Translation describes efforts to create a network where all the actors can agree that the network is worth building and sustaining. Callon (1986) postulated 4 elements of translation which include: Problematisation (identification of problems, relevant actors, and representatives), interessement (efforts to solicit actor interest and negotiate actor involvement), enrolment (actor commitments), and mobilization of allies (ensuring adequate actor representation of all groups involved in the network). Actants refer to the human and non-human actors that make up and shape the network through their interrelations within it. The stability and social order of networks depends on successful interactions within the actor-networks and the creation of tokens, quasi-objects exchanged between actors within the network. Major strengths include the ability to provide rich data on internal organizational processes and change. Major limitations include, lack of specificity concerning actant relations, ways to improve actant relations within the network relations, and an inability to account for non-actant influences from outside the network.

ANT limitations are offset by turning to complimentary perspectives like that focus on specific areas of actor-network relations such as the adoption of conversation theory for leveraging successful actant relations. Conversation theory provides a complimentary set of tools for promoting successful interactions within actor-networks known as learning conversations (grounded conversations). Luppicini, (2008) describes grounded conversation as "a complex multi-level conversation where shared understanding emerges from the co-construction of meaning embedded within multi-level *conversational processes* between two or more agreement seeking *conversational agents* within a specified social context (p.205). Conversational agents (similar to ANT actants) situated within a social content (in this case an actor-network) must engage in conversations to reproduce organization memory, participation in decision making, and build mutually shared knowledge for future network use. It is worth noting that there are other theories that bear on globalization and technology that are not covered here such as technology oriented theories that focus on the connection between specific aspect of media and social processes including, social presence theory (Short, et al. 1976) media richness theory (Daft & Lengel 1986), media synchronicity theory (Dennis & Valacich 1999), social identity model of deindividuation effects (Postmes, Spears, & Lea 1999.

Diffusion of Innovation Theory

Diffusion of innovation refers to a theory introduced by Everett Rogers to explain the at which rate new innovations (objects, behaviors, ideas) are spread through cultures. Rogers (2003) argued that diffusion was the process by which an innovation is spread over time among members within social system and that this spread follows an s curve pattern from innovators to early adopters to early majority, to late majority to laggards, until all but the persistent skeptics adopt the innovation . Rogers also believed that that this S curve was a standardized pattern with the following distribution by percentages: Innovators (2.5%) early adopters (13.5%), early majority (34%), late majority (34%), and laggards (16%). Rogers also believed that individuals' attitudes toward a new innovation was fundamental in its diffusion and the decision to adopt a new innovation was a process that occurs over time and passed through five stages: (1) Knowledge (acquiring knowledge about the innovation), (2) Persuasion (developing an opinion about the innovation), (3) Decision (deciding to accept or reject an innovation), (4), Implementation (put the new technology into practice, and (5) Confirmation (approving the decision to accept or reject an innovation. One of the strengths of this theory lies its its widespread application and ability to provide measurable data of almost any context. The major limitation is that diffusion of innovation research often relies heavily on macro-level quantitative analysis which does not get at other factors (personal, social, political, cultural) that influence the innovation process. The next section deals with technoethical theory as a middle range theory that appears to be compatible with theories covered in this section but with a focus on the ethical side of technology and globalization.

TECHNOLOGY AND GLOBALIZATION

A relational systems view of technology provides a useful conceptual framework for understanding globalization. The main reasons to frame globalization within a relational system of technology are threefold: (1) The conceptualization of technology as a relational system acknowledges the complex reality of multiple factors shaping globalization (system view) along with the interconnectedness (relational nature) of technology and human activity, (2) Technology has been a core driver of multiple dimensions of global development (economic, political, social, cultural, communications, education, etc.), (3) A relational systems view of technology includes material, along with organizational principles processes, which allow multiple dimensions of globalization to be examined at multiple levels (local, national, international).

This work treats globalization as a technological phenomenon driven by technological innovation within human activity. It places technology at the core of globalization as the driving force for the developing processes of globalization (and deglobalization). Technological relations adjust the organization and scope of developing processes within the interconnections of societies, cultures, institutions, and individuals across the world. In this sense. Globalization and deglobalization are

Figure 1. Relational system of technology and globalization

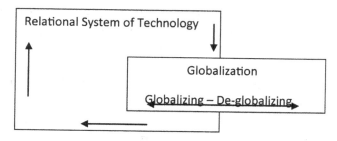

perhaps best represented as extreme ends of a continuum situated within a relational system of technology. To illustrate (see Figure 1).

How are technological relations and patterns of global interconnectedness organized and reproduced? In other words, what drives globalization within our technological society? There are complex set of sources driving global technological change that go beyond rationalistic explanations of technological progress. These sources, which shape and are shaped by, technological globalization are connected to human and social factors which affect the speed of R&D, the movements of market and leftover economies, and the role of political interests.

First, the speed of R&D can affect a great deal about how a potential service or product has a global impact. Proper testing and development of prototypes helps to safeguard new innovations, reduce risk, and ensure that what is being developed is in line with the needs and interests of those affected by it. This is especially true for controversial services and products that present a high level of potential of health and safety risk (i.e., genetically modified foods, nuclear power, etc.). The problem is that this requires a great deal of resources and time. Volti (2006) states, " The development work that goes into preparing a new technology for actual production can entail massive expenditures for equipment, material, manpower, pilot plants, and the like" (p.38). This may motivate researchers and technologists to take short cuts to reduce the investment of time and money.

Market economies are a second source driving global technological change. Volti (2006) reasons, "For a technology to make a transition from the potential to the actual requires not just that it exist; there must always be the desire for it coupled with the ability to pay for it (p.42)" Effective demand can be created by the presence of a market economy where there is an obvious interest and demand (i.e., mobile phone). This can also pave the way for continued effort to build on technological innovation already proven. As noted by Volti (2006):

The commercial viability for a new technology may thus stimulate the development of complementary technologies. A bottleneck that restricts the continued development of a particular technology creates strong economic incentives to find new technologies that clear the obstruction" (p.40).

It is worth noting that, in addition to market economies, there are also secondary and belated economies to consider. In the case that the effective demand of a product or service declines a global market place often provides a secondary marketplace. For instance, in the area of agriculture, a surplus of genetically modified food in the U.S. due to a loss in effective demand could be alleviated through increased export. For instance, genetically modified rice has been exported to India as part of U.S imposed foreign aid agreement. Other products banned in the U.S. such as DEET have been sold on the international market within nations where such a ban does not exist.

Finally, people and politics also are huge drivers of the global economy. Corporate forms led by a powerful elites and government institutions play a key role. Because of the large infrastructure needs for developing technological innovations, multi-national corporations with a great deal of capital have played a key role. Companies like Xerox, IBM, Mitsubishi, and Ford have their own R&D facilities for developing new products and services for the global marketplace. In addition, government institutions have also spearheaded many major initiatives that have impacted the global marketplace directly and/or indirectly. Take for example, NASA. The space shuttle program stimulated valuable research activities in space transport, telecommunications, and computer engineering which has employed many and many and placed man on the moon. At the same time, this technology has been reappropriated by other nations for other purposes. Canadian partnering with NASA allowed Canada to pioneer non-commercial communication satellite technology for leveraging Canadian broadcasting and communication abilities to rural regions I the north. This has changed the face of Canadian communications although much of the source of this innovation came from U.S. developed technology.

What are the Dimensions of Globalization?

Based on the above discussion, it can be seen that globalization is a multi-dimensional set of interconnected processes with many sources shaping their development and vice versa. In order to successfully study globalization within this complex arena of interrelational processes, theoretical frameworks must be employed to help explain key drivers, processes, players, and consequences of globalization. Volti (2006) argues, "Technologies are social creations, and any successful attempt at understanding why particular technologies are created, chosen, and used, must take into account their social context" (p.51). Perhaps the more effective way of capturing globalization is by understanding its many parts and exploring how these parts fit together within the relational system of technology. Based on an extensive review of the literature on globalization, key dimensions of globalization can be identified and placed within the taxonomy illustrated in Table 1 below.

These abovementioned dimensions represent taxonomy of interrelated areas of globalization discussed in multidisciplinary body of work on the subject. These dimensions are presented separately for clarity but are viewed as interrelated aspects of globalization. Social globalism affects the consciousness of individuals (individual globalism) as well as their held attitudes and cultural beliefs (cultural globalism). These dimensions of globalization are developing processes which can increase or decrease international flows (e.g., resources, ideas, services, products). Conversely, globalization also includes deglobalization which describes the flow from international interconnectedness to international separation and local or national interdependencies over time. The dimensions of globalization (and deglobalization) are elaborated on below.

First, theoretical globalism provides valuable conceptual perspectives with which to view globalization within the real world of rapidly expanding technology. There are a number of theories relevant to globalization and help frame research into different dimensions or areas of globalization. Among the most important theories are the follow: Global Polity, Global World, Social Constructivism, Actor-Network Theory (ANT), and Diffusion of Innovation. Taken together, there are a variety of theories available for explaining globalization processes, each with their strengths and weaknesses. Some theories focus exclusively on globalization but reduce globalization to a single dimension (General theories of globalization), while other theories are more open in terms of content but fail to explain specific dimensions of globalization connected to technology (social constructivist theories) or social context (technological theories). While some theories lend themselves easily to empirical research (diffusion of innovation), other theories do a much better at providing rich and detailed data that is context specific (ANT). None, of the theories address all moral and ethical considerations about globalization and technology. For this reason, a multi-perspective approach to global technoethics is recommended.

Second, economic globalism is an important dimension of globalization oriented to towards the new knowledge economy and rapid information exchange. Information technology lies at the core of modern globalization because it has provided the groundwork for the emerging knowledge-based economy, not to mention providing the underlying conditions which have fostered other developing processes and complex interconnections between societies, cultures, institutions, and individuals at a global level. Mowlana's (1997)*Global Information and World Communication: New Frontiers in International Relations* situates the information revolution

Table 1. Key dimensions of globalization and current trends

Dimension	Description	Current Trends
Theoretical globalism	This refers to a set of multidisciplinary theories focuses on the developing processes of globalization and its consequences.	Technocracy, mediation theory, global polity theory
Communication globalism	This refers to the international exchange of information and communication. It also involves the organization and processes linked to this exchange.	Global communications, virtual organizations, ICT's, Internet development
Economic globalism	This refers to the international flow of goods, services, and capital which constitute a global economy. It also involves the organization and processes linked to this flow.	Knowledge economy, telework, contingency workers, outsourcing
Environmental globalism	This refers to the international transformation of materials within the natural world which affect the biosphere, human life and other living organisms	Oil Tar sands, sour gas, ozone destruction, environmental contamination
Health and Medical globalism	This refers to the international flow of medical knowledge, services and products which affect human health and well-being. It also involves the organization and processes linked to this flow.	International tissue market, organs, tele-health, reproductive technologies
Educational globalism	This refers to the international development of education and training. It also involves the organization and processes linked to such developments.	Global accessibility to education, International DE, UNESCO,
Political globalism	This refers to the international networks of interdependence in which policy making, negotiation, and governance are employed. It also involves the organization and processes linked to these networks.	Global governance, international governing bodies, multinational corps., virtual alliances
Cultural globalism	This refers to the international movement of cultural values, practices, images, languages, and people. It also involves the organization and processes linked to these movements.	Online Cultural and Religious Communities, migration, multiculturalism
Social globalism	This refers to the international movement of social values, knowledge, practices, and institutions. It also involves the organization and processes linked to these movements.	Cyber Democracy, global citizenship, Human Rights movements, Neo-luddism
Individual globalism	This refers to the global consciousness of individuals, their sense of global identity, and personal attitudes towards globalization.	Global consciousness, identity
Ethical globalism	This refers to global ethical implications and dilemmas that affect the globalized world. This includes the organization and processes connected to these ethical implications and dilemmas.	Global Ethics, corporate responsibility, trust,

within the context of new technologies (mass communications, telecommunication, and new media) and at the hub of global economic expansion. Mowlana (1997) situates information technology as the driving force in the global economy as an information-based economy:

The global economy is now truly developing into an information-based economy. Such development and its effects include: the increasing flow of information and information-based products and services among nations; the growing economic importance of information and related products and services within and between nations; the increasing cultural and political significance of information and related products and services that do not correspond to traditional categories, the increased difficulty of enforcing intellectual property rights on the international level; and the growing convergence of international intellectual property issues with other national issues (p. 15)

The information revolution affected many areas of life and society while raising political, economic, cultural, technological, military, and legal issues connected to its developments. Mowlana's argues for the important role that information plays in world politics, economic advancement, cultural sustainability, environmental change, and various areas of international development. The information revolution has contributed to the creation of new communication technologies and systems which have changed many aspects of life around the globe. For this reason, Mowlana stresses the need to look at many areas of human activity to appreciate the global nature of information and communication. As stated by Mowlana (1997) "It takes an integrative approach to international communication by examining both the human and technological dimensions of global information (p. xi)

Third, the information revolution, not only paved the way for the knowledge-based economy, but also provided the conditions for global communication building. Mody's (2003) International and Development Communication: A 21st-century Perspective is an edited edition focusing on international (global) communication and development issues. This work also placed telecommunications at the center of globalization and deals with a variety of key issues including, the role of media, biases in media and telecommunication institutions, and power relations in international communications. As indicated by Boyd-Barrett's article, entitled, "Global Communication Orders" This order is adjudicated and adjusted by "global" institutions such as the World Trade Organization, the World Bank, and the International Monetary Fund, in which the wealthy nations exercise most influence." (p. 36). Taken together, this work helps to ground globalization within information and communication developments arising from a relational network of technology

Fourth, the technological driving force in globalization is also recognized in scholarship work on globalization and culture. Tomlinson's (1999)*Globalization and Culture* assumes "that the huge transformative processes of our time that globalization describes cannot be properly understood until they are grasped through the conceptual vocabulary of culture." In doing so, the author picks up on the current strain placed on many cultures due to increased exposure to global trade structure and way of life:

The cheap prices of its commodities are the heavy artillery with which it batters down all Chinese walls, with which it forces the barbarians' intensely obstinate hatred of foreigners to capitulate. It compels all nations, on pain of extinction, to adopt the bourgeois mode of production; it compels them to introduce what it calls civilization into their midst, i.e., to become bourgeois themselves (p.76)

This work is important in further extending understanding of globalization processes and raise awareness of the complexity of how global interactions affect local contexts and alter the construction of meaning in everyday life and culture.

Fifth, global education is another important dimension in the technology driven world. Pervasively high migration rates between nations, new economic realities, new technologies, increasing cultural diversity in urban centers around the world, and increased influences in education from the corporate sector is providing educators with a number of challenges and opportunities for global education. Stromquist's (2002)*Education in a Globalized World: The Connectivity of Economic Power, Technology, and Knowledge* provides a forward thinking examination of key challenges in global education under the influence of technology driven economic trends and demands for education and training intervention. Stromquist (2002) draws attention to the immense challenge in leveling the playing field in skills and training needed for success in a global economy:

The process that has come to be called "globalization" is exposing a deep fault line between groups who have the skills and mobility to flourish in global markets and those who either don't have these advantages or perceive the expansion of unregulated markets as inimical to social stability and deeply held norms (p. 92).

Another excellent work comes from the edited collection of essays within Suárez-Orozco's (2007)*Learning in the Global Era: International Perspectives on Globalization and Education. The collection of essays provides an interesting set of articles exploring key challenges in global education including, the education of* youth within a technologically advanced and interconnected world, the strategies used by educational institutions to meet the educational needs of immigrant youths at risk, strategies used by educational institutions and other organizations to provide education and training to overcome problems of access and inequality, and efforts to transcend boundaries of cultural tolerance and acceptance of diversity in educational institutions schools.

Fifth, social globalism is an important dimension of globalization being transformed by the presence of new technologies and social relations. Given that global citizenship is a desired end, what means might lead to its realization? One option to adapt an ideological stance that commits citizens to uphold certain commitments to society. For instance, Findlay (2000) views global citizenship situated in a communitarian society which reinforces individual interrelations and cultural commitment:

For a society to be communitarian, its heavily enmeshed fabric of interdependencies therefore must have a special kind of symbolic significance to the populace. Interdependencies must be attachments which invoke personal obligation to others within a community of concern. They are not perceived as isolated exchange relationships of convenience but as matters of profound group obligation. Thus, a communitarian society combines a dense network of individual interdependencies with strong cultural commitments (p. 31)

Sixth, political globalism is an important dimension of globalization embedded in economic structures and public institutions. Siebert's (2002)*Global Governance: An Architecture for the World Economy* provides an excellent look at how political globalism influences the economy and raises public concern about how governments should deal with globalization challenges. Siebert (2002) explores way to improve the role of public institutions (national and international) in responsible global governance. Some key suggestions revolve around promoting international trade, guarding against international financial crises, and ensuring the sustainability of global environmental systems. This dimension of globalism provides an important look at globalization connected to issues of responsible governance for global citizens.

Seventh, environmental globalism has gradually become a central dimension of globalization focus. Many example of technology driven environmental disasters (strip mining, oil spills, chemical leaks) and ongoing threats (i.e., mass terrorism, bio-warfare, nuclear weapons) have led environmental globalism to become a major area of concern in globalization debates interconnected with other dimensions. Al Gore's (2007) Inconvenient Truth is one example of how the global environment is changing rapidly within a technology driven global economy. Although future technologies (e.g., more energy efficient fuel sources) may help the problem, Volti (2006) notes that some new technology cannot solve the problems of pollution and climate change. This is because technological change is influenced by social, political and cultural factors that must be taken into consideration in any efforts to solve problems caused by technology such as government policy implementation on industries (p.104).

Eight, health and medical globalism is another important dimension of globalization. This is especially salient in areas where medical research spearheaded by developed countries draws on the population of developing countries. Although there are many positive global initiatives in health and medicine, this can also lead to the exploitation of developing countries and risk of injury for their citizens. For instance, in the mid-1990s Nigeria has a major epidemic of meningococcal meningitis (or cerebrospinal meningitis,) killing over 10,000 people (Perlroth, 2008). Perlroth (2008) noted how Pfizer (a major drug company) began conducting clinical trials in Nigeria on a new experimental drug (Trovan) which violated major ethical research protocols which led to deaths and injuries of several test subjects. Situations like this call for increased attention to how health and medical globalism is developed and implemented.

Ninth, individual globalism and global consciousness I a relatively new dimension of modern globalization that is beginning to gather momentum. Becoming a so called 'world citizen' with a global consciousness describes an ideological perspective on what globalization could offer individuals. Findlay (2000) states, "Globalization as a concept refers both to the compression of the world and the intensification of consciousness of the world as a whole . . . both concrete global interdependence and consciousness of the global whole (p.3). It has been argued that one of the key consequences of the information revolution was to leverage global communications and global consciousness through opportunities for increased awareness of the global condition (Giddens 1991, 1991; Albrow 1996). This dimension of globalization is a promising area of current and future inquiry.

Tenth, ethical globalism has moved from being a secondary dimension of globalization to being a core dimension and important concern within the globalization discourse. Hemelink (1997) provided insight about how a new paradigm of global morality can be achieved by national around the world by relying on current information and communication technologies available today. Hemelink states,"Make it their aim to facilitate the freer and wider dissemination of information of all kinds, to encourage co-operation in the field of information and the exchange of information with other countries" (p. 105). Not only does Hemelink view global morality as an achievable goal, but also treats it as a means for achieving a greater moral aim within a globalizing world:

The right to development is an inalienable human right by virtue of which every human person and all peoples are entitled to participate in, contribute to and enjoy economic, social, cultural and political development, in which all human rights and fundamental freedoms can be fully realized (p.108)

The ethical dimension of globalization is particularly important within the current globalization context defined oriented towards rapid technological progress. Ethical consideration help to establish rules and policies to help better control and direct globalization efforts to provide global citizens adequate protection from exploitation and unnecessary risk as globalization advances.

Taken together, these ten dimensions of globalization provide a broad view of the breadth and complexity of modern globalization today within a rapidly expanding technological world. These technology oriented dimensions of modern globalization form the basis of global technoethics.

DISCUSSION QUESTIONS

- To what extent does the global knowledge-based economy influence the level global ethical dilemmas?
- How can your country become a leader in imposing strict regulations on industry ethical standards? How can regulations be imposed without sacrificing economic opportunities?
- If the local development of technologies is valued over global partnerships, what might be the consequences for developing countries?
- Many experts have drawn attention to the global digital divide in technology access? How do you think this has influenced developed and developing nations?
- of the audience through photos, video, or digital representations (i.e., avatars)?
- Is global citizenship possible without shared geographic space and shared local cultural meaning?
- How should one use such images in a manner that is ethically correct, i.e. that does not mislead people? Various
- If a code of ethics is developed by an organization fin one country, can it be applied outside of the culture in which it was developed?
- What ethical issues concern multinational organizations with cutting-edge technology?

DISCUSSION

This chapter introduced global technoethics as an important area of inquiry focused on the ethical use of technology to promote the aims of globalization. It described a systems view of technology to help situate globalization within the complex reality of technology and human activity shaped by interconnected influences. In articulating global technoethics, this chapter placed technology and human activity at the core of globalization. Global technoethics extends the analysis of technology and ethics to a global scale dealing with questions such as technology and developing nations, cultural assimilation and exploitation of foreign populations and resources.

REFERENCES

Austen, I. (June 20, 2008). 3 Ex-Nortel executives are accused of fraud. *The New York Times*. Retrieved February 1, 2009 from http://www.nytimes.com/2008/06/20/technology/20nortel.html

Bijker, W., Hughes, S., & Pinch, T. (1987). *The social construction of technological systems, new directions in the sociology and history of technology.* Cambridge MA: MIT Press.

Boyd-Barrett, O. (1997). Global communication orders. In B. Mody (Ed.), *International and development communication: A 21st-century perspective (pp.35-52).* Thousand Oaks, CA: Sage.

Callon, M. (1986). Some elements of a sociology of translation: domestication of the scallops and the fishermen of St. Brieuc Bay. In J. Law (ed.), *Power, action and belief: A new sociology of knowledge.* London: Routledge & Kegan Paul.

Corpwatch. (2009). Retrieved June 1, 2009 from http://www.corpwatch.org/

Corpwatch (March 20, 2003). *Halliburton makes a killing on Iraq war.* Retrieved February 1 2009, from http://www.corpwatch.org/article.php?id=6008

Daft, R. L., & Lengel, R. H. (1986). Organizational information requirements, media richness and structural design. *Management Science, 32*(5), 554–571. doi:10.1287/mnsc.32.5.554

Dalkir, K. (2005). *Knowledge management in theory and practice.* Burlington, MA: Elsevier Butterworth-Heinemann.

Denis, A., & Valacich, J. (1999). Rethinking media richness: towards a theory of media synchronicity. In *Proceedings of the 32nd Hawaii International Conference on Systems Science.*

Desanctis, G., & Poole, M. S. (1994). Capturing the complexity in advanced technology use: adaptive structuration theory. *Organization Science, 5*(2), 121–147. doi:10.1287/orsc.5.2.121

Dewey, J. (1927). *The public and its problems.* New York: Holt.

Giddens, A. (1984). *The constitution of society: outline of the theory of structuration.* Berkley, CA: University of California Press.

Goold, D., & Willis, A. (1997). *The Bre-X fraud.* Toronto: McClelland and Stewart.

Gore, A. (2007). *An inconvenient truth: The crisis of global warming.* New York: Viking/Rodale.

Haag, S., Cummings, M., McCubbrey, D., Pinsonneault, A., & Donovan, R. (2006). *Management information systems for the information age* (3rd Canadian Ed.). Canada: McGraw Hill Ryerson.

Hemelink, C. (1997). The new paradigm and global morality. In A. Mohammadi (Ed.), *International communication and globalization: A critical introduction (pp. 90-118).* Thousand Oaks, CA: Sage.

James, J. (2006). *Globalization, information technology and development.* New York: St. Martin's Press.

Jeter, L. (2003). *Disconnected: Deceit and Betrayal at WorldCom.* New York: Wiley.

Kaplan, D. (September 18, 2006). Suspicions and spies in silicon valley. *Newsweek.* Retrieved February 1, 2009 from http://www.newsweek.com/id/45548/

Knorr-Cetina, K. (1999). *Epistemic cultures: How the sciences make knowledge.* Cambridge, MA: Harvard University Press.

Kochan, N., & Pym, H. (1987). *The guinness affair: Anatomy of a scandal.* NY: Helm Information.

Latour, B., & Woolgar, S. (1986). *Laboratory life: The construction of scientific facts.* Princeton, NJ: Princeton University Press.

MacLean, B., & Elkind, P. (2002). *Smartest guys in the room: The amazing rise and scandalous fall of Enron.* New York: Routledge.

Merz, M. (2006). Embedding digital infrastructures in epistemic cultures. In C. Hines (ed.), *New infrastructures for knowledge production: understanding e-science (pp. 100-115).* Hershey, PA: Idea Group.

Meyer, J., Boli, J., Thomas, G., & Ramirez, F. (1997). World society and the nation-state. *American Journal of Sociology, 103*(1), 144–181. doi:10.1086/231174

Miller, K. (2008). *Organizational communication: Approaches and processes* (5th Ed.). Belmont, CA: Thomson Wadsworth.

Modelski, C. (1972). *Principles of world politics.* New York: Free Press.

Mody, B. (2003). *International and development communication: A 21st-century perspective.* Thousand Oaks, CA: Sage.

Mowlana, H. (1997). *Global information and world communication: New frontiers in international relations.* Thousand Oaks: Sage.

Nentwich, M. (2003). *Cyberscience: Research in the age of the Internet.* Vienna: Austrian Academy of Science Press.

Perlroth, N. (December 8, 2008). Pfizer's Nigerian nightmare. *Forbes.* Retrieved February 1, 2009, from http://www.forbes.com/forbes/2008/1208/066.html

Perlroth, N. (January 30, 2009). Pfizer's Nigerian plaintiffs get day in court. *Forbes.* Retrieved February 1, 2009, from: http://www.forbes.com/2009/01/30/pfizer-nigeria-trovan-business-healthcare-0130_trovan.html

Pinch, T., & Bijker, W. (1992). The social construction of facts and artifacts: or how the sociology of science and the sociology of technology might benefit each other. In W. Bijker & J. Law (eds.), *Shaping technology/building society (pp. 17-50).* Cambridge, MA: MIT Press.

Postmes, T., Spears, R., & Lea, M. (1999). Social identity, group norms, and deindividuation: Lessons from computer-mediated communication for social influence in the group. In N. Ellemers, R. Spears (eds.), *Social identity: Context, commitment, content.* Oxford: Blackwell.

Robbins, S., & Judge, T. (2007). *Organization Behavior (12th ed.).* Upper Saddle River, NJ: Pearson Prentice Hall.

Robert, D., & Backes, E. (2002). *The silence of the money. The Clearstream scandal.* New York: Routledge.

Rogers, E. (2003). *Diffusion of innovations* (5th Ed.). New York: Free Press.

Salinger, L. (2004). *Encyclopedia of white-collar and corporate crime.* Thousand Oaks, CA: Sage.

Schilling, G. (2000). The virtual observatory moves closer to reality. *Science, 289,* 238–239. doi:10.1126/science.289.5476.29

Seglin, J. (2003). *The right thing: Conscience, profit and personal responsibility in today's business.* Spiro Press.

Short, J. A., Williams, E., & Christie, B. (1976). *The social psychology of telecommunications.* New York: John Wiley & Sons.

Siebert, H. (2002). *Global governance: An architecture for the world economy.* Berlin: Springer.

Stromquist, N. (2002). *Education in a globalized world: The connectivity of economic power, technology, and knowledge.* Lanham, MD: Rowman and Littlefield.

Suárez-Orozco, M. (Ed.). (2007). Learning in the global era: International perspectives on globalization and education. Berkeley, CA: University of California Press.

Today, U. S. A. (June 17, 2005). *Timeline of the Tyco International scandal.* Retrieved February 1, 2009 from http://www.usatoday.com/money/industries/manufacturing/2005-06-17-tyco-timeline_x.htm

Tomlinson, J. (1999). *Globalization and culture.* Chicago: University of Chicago Press

UNCTAD. (1996) *World Investment Report.* UNCTAD.

United Nations Mission in Liberia. (2007). *Human rights in Liberia's rubber plantations: Tapping into the future* (pp. 45-46). United Nations Mission in Liberia. Retrieved February 1, 2009 from, http://www.stopfirestone.org/liberiarubber.pdf#page45

Volti, R. (2006). *Society and technological change,* (5th Ed.). New York: Worth Publishers.

Wallerstein, I. (1974). The rise and future demise of the of the world-capitalist system: concepts for comparative analysis. *Comparative Studies in Society and History, 16,* 387–415. doi:10.1017/S0010417500007520

Wallerstein, I. (1998). *Utopistics: Or, historical choices of the twenty-first century.* New York: The New Press.

Wouters, P., & Schroder, P. (Eds.). (2003). *Promise and practice in data sharing.* Amsterdam: NIWI-KNAW.

York, B. (July 14, 2003). Halliburton: The Bush/Iraq scandal that wasn't. *National Review.* Retrieved February 1, 2009 from http://www.nationalreview.com/york/york070903.asp

ADDITIONAL READING

Boli, J., & Thomas, G. (1997). World culture in the world polity. *American Sociological Review, 62*(2), 171–190. doi:10.2307/2657298

Castells, M. (1999). *Information technology, globalization and social development.* Geneva: United Nations Research Institute for Social Development.

Freeman, C., & Hagedoorn, J. (1995). Convergence and divergence in the internationalization of technology. In J. Hagedoorn (ed.), Technical change and the world economy (pp. 214-225). Aldershot: Edward Elgar.

George, R. (2006). Information technology, globalization and ethics. *Ethics and Information Technology, 8*(1), 29–40. doi:10.1007/s10676-006-9104-4

Hanna, N., Boyson, S., & Gunaratne, S. (1996) The East Asian miracle and information technology. World Bank Discussion Papers, no. 326.

Latour, B. (1987). *Science in action: How to follow scientists and engineers through society.* Milton Keynes: Open University Press.

Latour, B. (2005). Reassembling the social: an introduction to actor-network-theory. Oxford: Oxford University Press.

Law, J., & Hassard, J. (Eds.). (1999). *Actor network theory and after.* Oxford: Oxford and Keele.

McMahon, P. (2002). *Global control: Information technology and globalization since 1845.* Cheltenham, UK: Edward Elgar.

Meyer, J. W. 1980. The World Polity and the Authority of the Nation-State. In A. Bergesen (ed.), *Studies of the modern world-system (pp. 109-137).* New York: Academic Press.

Robertson, R. 1991. "The Globalization Paradigm: Thinking Globally." Pp. 207-24 in Religion and Social Order. Greenwich: JAI Press.

Rosenau, J., & Singh, J. (2002). *Information technologies and global politics: The changing scope of power and governance.* Albany: State University of New York Press.

Stromquist, N., & Monkman, K. (2000). *Globalization and education: Integration and contestation across cultures.* Lanham, MD: Rowman and Littlefield.

Chapter 11
The Case of Global Technology in South Africa

INTRODUCTION

Perhaps the greatest challenge 21st century globalization i to overcome the technological divide (this includes the digital and non-digital technologies) that exists between
developed and developing nations. The uneven global uptake of technology is one
of most important gauges of global inequality in the world today. Despite substantive efforts to level the playing field by bringing new opportunities to developing
nations, this challenge continues to plague out modern world. It touches every area
of human activity in society that depends on technology and change. It is both important in the current context and in future technology development. This raises the
question about how it is that technology diffuses at a global level and how should
this diffusion be regulated and controlled. For instance, Jeffrey (2001) examines

DOI: 10.4018/978-1-60566-952-6.ch011

the economic characteristics of ICT's to gauge their potential effects on the global economy. Jeffrey (2001) found that ICT's "are associated with a number of powerful cumulative mechanisms causing some countries to grow rapidly and others to become increasingly marginalized from the global economy (p. 147). According to Jeffrey, South Africa is among the most marginalized of developing nations despite substantive investment in their ICT infrastructure from the international development initiatives. Why is this the case and what can be done to improve the situation?

Global technoethics is concerned with the ethical use of technology to promote the aims of globalization. Because of the centrality of information and communication technology in the world economy and its capacity to create worldwide linkages at all levels of globalization processes currently studied (i.e., economic, technological, political, cultural, environmental), global technoethics places globalization within a technological relational system. In order to understand global technoethics and its unique contribution, it is necessary to situate it within the existing globalization literature. This chapter focuses on recent technological transformation current underway in South Africa (SA). Due to the complexity and extent of the ongoing challenges, a multi-perspective examination of key technology innovation development efforts is used to explore possible weaknesses and imbalances blocking efforts to leverage South Africa.

BACKGROUND

What is globalization? How does globalization affect individuals and societies? Does globalization lead to poverty or prosperity? Why are so many debates surrounding globalization and how it is measured? Does globalization diminish cultural diversity or enhance it? Can globalization be controlled and by whom? These are some of the questions at the heart of the globalization debate. But, what is globalization? A variety of different definitions for globalization have been posited over the years with no single definition receiving widespread acceptance. The following are among the most prominent definitions of globalization:

- "The historical transformation constituted by the sum of particular forms and instances of . . . . [m]aking or being made global (i) by the active dissemination of practices, values, technology and other human products throughout the globe (ii) when global practices and so on exercise an increasing influence over people's lives (iii) when the globe serves as a focus for, or a premise in shaping, human activities" (Albrow, *The Global Age*, 1996, p. 88).
- "Globalization denotes the expanding scale, growing magnitude, speeding up and deepening impact of interregional flows and patterns of social interaction.

It refers to a shift or transformation in the scale of human social organization that links distant communities and expands the reach of power relations across the world's major regions and continents(Held, 2003, p.2)"

- Globalization as a concept refers both to the compression of the world and the intensification of consciousness of the world as a whole . . . both concrete global interdependence and consciousness of the global whole " (Findlay, 2000, p.3)

- "[T]he inexorable integration of markets, nation-states, and technologies to a degree never witnessed before-in a way that is enabling individuals, corporations and nation-states to reach around the world farther, faster, deeper and cheaper than ever before . . . . the spread of free-market capitalism to virtually every country in the world " (T.L. Friedman, *The Lexus and the Olive Tree*, 1999, p. 7-8).

- "Globalization can thus be defined as the intensification of worldwide social relations which link distant localities in such a way that local happenings are shaped by events occurring many miles away and vice versa (Giddens, 1991, p. 64)

- "The rapidly developing processes of complex interconnections between societies, cultures, institutions, and individuals world-wide" (Miller, 2008).

- The compression of the world and the intensification of consciousness of the world as a whole . . . . concrete global interdependence and consciousness of the global whole in the twentieth century" (Robertson, *Globalization*, 1992, p. 8).

- "A social process in which the constraints of geography on social and cultural arrangements recede and in which people become increasingly aware that they are receding" (Waters, *Globalization*, 1995, p. 3).

What is interesting to note is the uniqueness of each definition in terms of what processes are considered to globalizing processes and what is the scope of these globalizing processes. On the one hand, the uniqueness of definitions draws attention to the richness of perspectives and broad scholarly interest in this area. On the other hand, the lack of a single accepted definition creates ambiguity and opens the door for misinterpretation and confusion.

Why is the globalization challenging to grasp? One of the main obstacles in understanding globalization concerns the ambiguity surrounding how it is defined and operationalized. Held & McGrew (2003) state, "Trying to make sense of this debate presents some difficulties, since there are no definitive or fixed lines of contestation. Instead, multiple conversations coexist (although few real dialogues), which do not readily afford a coherent or definitive characterization" (p.2). How can any strong support for or against globalization belevied when there is little agreement about what it is? The term, "globalization" has been used in different ways to describe social, political, cultural, and economic developments emerging from their growing interconnectedness. Anthropologists and sociologists typically focus on cultural changes and community practices, while economists typically focus on finances and the exchange of goods and services within a global market. Political scientists typically focus on the changing power of transnational corporations and the role of international non-governmental organizations and agencies (i.e., World Bank, United Nations), while communication experts typically focus on the spread of information and communication technologies and cultural change. Therefore, globalization in communications (global village perspective) does not mean the same as globalization in economics (global economy), which does not mean the same as globalization in political science (global citizen perspective). The question arises as to which perspective or combination of perspectives on globalization is one to choose? Should different perspectives on globalization be treated as inter-related parts of the same phenomenon or as different phenomena altogether? Does the globalization of the economic processes entail the globalization of culture and identity? If so, under what conditions? Does the globalization of the culture entail the globalization of economic processes? If so, under what conditions? Does the globalization of the economic processes entail the globalization of politics and society? If so, under what conditions?

One approach to overcoming the conceptual disconnect in globalization discourse is to choose one perspective and to operationalize it in an effort to measure changes of globalization. For instance, anthropologists and sociologists often measure migration patterns and examine the percentage of migrants residing in different regions and the flow of migrants between regions. Communication scholars often measure changes in global transportation and communication costs, changes in the digital divide, and the spread of technologies (e.g., Internet, mobile phone, PDA's etc,) across the world. This approach to conceptualizing globalization forces one to choose among the various measures. While this does allow for a clearer focus, it fails to other key dimensions of globalization.

Another approach, the one pursued in this work, views globalization as a set of interconnecting processes linked together within a larger system. This multidimensional conception of globalization is consistent with contemporary views of social reality as constituted by reflexivity and the interweaving of multiple elements (Giddens 1991). Giddens (1991) states, "The reflexivity of modern social life consists in the fact that social practices are constantly examined and reformed in the light of incoming information about those very practices, thus constitutively altering their character" (p.38). The current condition described by Giddens is driven by the advent of telecommunications which increased the speed, flexibility, and capacity of this reflexivity that characterizes modern social life.

The current work frames globalization within a technological system because of the current importance of communication and technology in the world economy and its capacity to create worldwide linkages at all levels of globalization processes currently studied (i.e., economic, technological, political, cultural, environmental). This follows Held and McGrew's (2003) view of technology as a revolutionary development of modern society within which globalization is situated:

Nations, peoples and organizations are linked, in addition, by many new forms of communication which range across borders. The revolution in micro-electronics, in information technology and in computers has established virtually instantaneous worldwide links, which, when combined with the technologies of the telephone, television, cable and satellite, have dramatically altered the nature of political communication. The intimate connection between 'physical setting', 'social situation' and politics, which distinguished most political associations from pre-modern to modern times, has been ruptured; the new communication systems create new experiences, new modes of understanding and new frames of political reference independently of direct contact with particular peoples, issues or events (p. 40).

Globalization in the 21st century is primarily situated within a technological system of relational patterns producing (and reproducing) increasingly complex interconnections of societies, cultures, institutions, and individuals across the world. As will be elaborated on below, the modern technological framework of globalization began materializing in the 1960s and 1970s with the advent of technology driven industries (i.e., biotechnology, computing, genetically modified foods), technological driven threats (i.e., mass terrorism, nuclear weapons, environmental disasters) and the establishment of a global telecommunications infrastructure that transformed information and communication exchange around the world.

World Bank's (1996) *Global Economic Prospects and the Developing Countries* found that the ratio of FDT/GDP in developing countries from 1990 to 1993 increased slightly overall but that over 75% of this increase was concentrates in 10 developing countries. At the same time, the trade ratios fell in 44 out of the 93 developing countries studied. Similar patterns were observed in foreign investment data (FDI/GDP). According to the World Bank (1996). 'Over the past decade ratios of FDI to GDP fell in thirty-seven of the ninety-three countries studied. Of these twenty were in Sub-Saharan Africa, nine were in Latin America and the Caribbean and seven were in the Middle East and North Africa' (p. 22)

Technology is considered to be a major tool in helping developing countries advance themselves through international trade development. Jeffrey (2001) views information and communication technology (ICT) as a major driver of international trade because of its role in changing transport costs and the costs of communication between buyers and sellers in different countries. According to Jeffreys:

There are two distinct mechanisms through which globalization is currently being driven by a reduction in communication costs. The first and more obvious mechanism is that because of technologies such as digital switching, fax machines and the Internet, information about already traded goods can be communicated between countries much more cheaply than was hitherto possible. . . . The second mechanism, by contrast, concerns the ability of information technology to bring certain services that were formerly non-tradable into the realm of tradability (p. 150)

In other words, technological differences are a major engine of trade because it provides some developing countries advantages that impact their international production capacities. Developing countries equipped with some level of industrial technology are leading those that do not. Moreover, the use of ICT's by multinational enterprises has also provided a push for globalization through increased ratios of foreign direct investment to gross domestic product. As noted by Dunning and Narula (1996):

It is a fundamental feature of MNE [multinational enterprise] activity that cross-border market failure exists in the supply of intermediate products, and especially intangible assets. ICT [information and computer technologies] has reduced both the costs of acquiring and disseminating information, and the transaction and co-ordination costs associated with cross-border activity. This is on at least two levels. First, information about both input and output markets is more easily accessible. This allows firms which previously could not engage in international business transactions now to do so. . . Second, MNEs are better able to integrate the activities of their various affiliates through the use of these technologies and to more quickly respond to changing conditions in the countries in which they operate (p. 8)

How do developing countries exploit information technology to integrate themselves more closely into the global economy? In drawing on examples of successful efforts to globalize, Jeffries (2001) identifies a number of ways that technology is used for promoting economic globalization in developing countries by reducing communication costs, enhancing comparative advantage of adopting firms, increasing electronics exports, and promoting strategic alliances by multinational corporations and other organizations. This suggests that developing countries like South Africa need to understand the complex nature of ICT globalization initiatives to capitalize on these opportunities. To this end, major international ICT development initiatives in South Africa are sketched out in order to identify key developments and challenges.

Table 1. Key services and initiatives

Name	Description	URL
E-Strategies - Empowering Development	This strategy of the International Telecommunication Union (ITU) has the goal to assist developing countries in harnessing the potentials of ICT to contribute towards reducing the social divide, improving the quality life, promoting universal access and facilitating entry into the information society.	http://www.digi-taldividenetwork.org/
Acacia Initiative	The Acacia Initiative: Communities and the Information Society in Africa Program Initiative is an international program to empower sub-Saharan communities with the ability to apply information and communication technologies (ICTs) to their own social and economic development.	
Africa service	Africa service offers information services and free computer training. It developed and deployed numerous software, fonts, multimedia applications and keyboards to compliment other efforts to ensure that Africa is not left behind in ICT.	http://www.africaservice.com
African Information Society Initiative (AISI)	The African Information Society Initiative (AISI) is an action framework that has been the basis for information and communication activities in Africa since 1996. AISI is not about technology. It is about giving Africans the means to improve the quality of their lives and fight against poverty.	http://www.uneca.org/aisi/

THE CASE OF GLOBAL TECHNOLOGY IN SOUTH AFRICA

Description

Post-apartheid South Africa has been faced with the challenge attempting to re-engineer its political and socio-economic structures to level racial inequalities, to eradicate poverty, to bridge the gap between the rich and the poor, to reduce regional imbalances, and to provide universal access to health care and education. This has proved to be a major challenge for South Africa as indicated by Baskaran, Muchie and Maharajh (2006), "As the legacy of apartheid lingers on, the country is facing many obstacles and the exclusion and division with South African society continue to be major problems" (p.181). A number of instrumental initiatives and organizations have been established to help with this difficult transition. These have been compiled from the UNESCO database of current development initiative and presented below in Table 1 and Table 2 along brief descriptions and contact information

E-Strategies - Empowering Development, Acacia Initiative, Africa service, African Information Society Initiative (AISI) are key service in place to assist South Africa in developing a technology infrastructure and the ability to use it to leverage education, job opportunities, communications, and economic development

African Telecommunications Union (ATU), Translate.org.za, South Africa National Information Technology Forum (NITF), Southern African Non-Governmental

Table 2. Key networks and organizations

Name	Description	URL
African Telecommunications Union (ATU)	ATU is a specialized agency of the African Union, in the field of telecommunications. Its mission is to promote the rapid development of info-communications in Africa in order to achieve universal access, and full inter-country connectivity.	http://www.trasa.org.bw
Translate.org.za	Translate.org.za is a non-profit organization producing free and open source software to empower South Africans. The Translate Project started in 2001 with the aim to provide free software translated into the 11 official languages of South Africa.	http://www.translate.org.za/
South Africa - National Information Technology Forum (NITF)	The NITF is committed to developing and advocating clear policy positions and policy options for the creation and advancement of the South African Information Society.	http://www.sn.apc.org/nitf/
Southern African Non-Governmental Organization Network (SANGONeT)	SANGONeT is a facilitator in the effective and empowering use of information and communication technology (ICT) tools by development and social justice actors in Africa. Its aims to share information, build capacity and link people and organisations through the use of ICTs.	http://www.sn.apc.org/corporate/index.shtml
Women's Net	Women's Net is a vibrant and innovative networking support program designed to enable South African women to use the Internet to find the people, issues, resources and tools needed for women's social activism.	http://womensnet.org.za/

Organisation Network (SANGONeT), Women's Net provide dedicated support for key areas of ICT development and integration in South African life and society. This includes the offering of services to promote communication and translation services, social justice, and increased technology access.

Despite what appears to be an excellent array of ICT development initiatives, a technoethical inquiry reveals a number of challenges that have plagued development efforts. Within a technoethical framework, technology relations to historical, social, political, cultural, communication, and educational factors help explain the persistence of barriers blocking efforts to create an effective national technology infrastructure to leverage South Africa. These are addressed within the following analysis.

Analysis

What are the historical relations blocking technological progress? South Africa has been an experiment in international development efforts for decade with uneven results still fresh in the minds of South African people. This was particularly devastating in the 1980s when South Africa agreed to an assisted restructuring of major infrastructures which did not improve conditions. Muchie and Baskaran (2006) noted the historical failures of global development efforts:

In the 1980s a large number of very poor and vulnerable economies in Africa tasted the bitter medicine of structural adjustment where the key capacitating institutions such as education, health and infrastructure were mostly crippled by public spending cuts ostensibly because it was believed that getting macro-economic balance between national revenue and public expenditure is considered more important than spending in building mental capital as a means of creating wealth. Many economies faced a double bind where they had to export based on what they extract and farm in order to pay debt and service debt payments. As a consequence many of these economies became failed economies with negative or stagnant rates of economic growth, hardly registering any new social economic development and failing educational, health and infrastructural foundations (p.241)

The wasted time and effort (not to mention the waste of limited resources) of past development initiatives has affected the confidence of South Africa in fully embracing new development initiatives. Assuming that successful development efforts require a high level of trust and collaboration between South African and leaders of development efforts, it is evident that the lack of both have played a role in holding back recent development efforts involving non South African led development initiatives.

What are the social and political relations blocking technological progress? Although Aparthied was legally abolished more than a decade ago, this does not eliminate the social and political inequalities built up over generations of slavery. Given that technology experts are highly educated and trained professional, the possibility of having a strong level of collaboration at all levels of a technology development initiative are slim. What makes this such a serious problem for current global development initiative is that the exclusion and division within South African society is entrenched within their social institutions and system of science and technology innovation. Mouton (2003) explains, "the production of science is still dominated by historically white universities and technikons, and the majority of scientists in the public sector are still white and male" (p.254). This means that South Africa is not able to take a leading role in their own technological infrastructure development and must rely on white expertise and foreign investors and stakeholders which do not fully understand the needs and interests of the South African people. Persisting social and political inequalities have contributed to an atmosphere or distrust and skepticism about the aims of external stakeholders in investing in the South African ICT infrastructure. This skepticism is shared by expert researchers working in this area. Muchie and Baskaran (2006) noted:

If ICT is used mainly by external corporations, governments and other bodies engaged in facilitating speed and movement of primary commodity exports that these economies largely specialize in on the basis of static comparative advantage, then ICT mainly will help in recording and storing data and accelerating the rate of circulation and transaction of already produced commodities p. 239).

What is interesting to note is the strong link between technoethics and the historical, social, and political relations deeply intertwined with South African ICT development efforts. The ethical dilemmas connected to human rights and inequality thought to be eliminated in 1989 are still deeply ingrained in the social institutions and lives of South Africa. The inability of ICT development efforts to address the pervasive inequalities and atmosphere of skepticism and distrust blocks the potential impact (and benefit) of technological innovations and support services available.

Solutions

From a technoethical standpoint, development experts need to take a step back to rebuild ethical relations and trust before moving forward with the development and implementation of technological innovations. First, following the logic of global technoethics (see Chapter 10), globalization is multidimensional and requires a relational approach. This is particularly evident in light of negative consequences experienced within poorer developing economies like South Africa. Even if development initiatives are clearly focused on one area (I.e. economy), development efforts must have a broader focus in order to leverage South Africa in becoming globally competitive. Second, globalization should begin at home. Globalization in South Africa involves humans with the right to participate and decide on the nature and direction of developments affecting them. In the context of South Africa, this means that educational technology programs and governing bodies for national level technology planning have to be geared towards empowering South Africans and including them at all levels. This is echoed by Muchie and Baskaran (2006) who state that "unless development of ICT is also shaped within a socially and politically framed national system of innovation, it is likely to create more diseconomies than economies "p. 241). Third, because of multidimensional nature of globalization a multi-faceted approach to ICT development is needed. In a case of South African ICT infrastructure development, this might entail, as suggested by Muchie and Baskaran (2006):

It is possible to approach the challenge by making national systems of innovation whilst learning and building the capacity for introducing new technologies through globalization and undertake policy learning to found an ICT sector however small that is, diversify the primary commodity economy and introduce new products and processes on the basis of knowledge-innovation, and extend ICT's benefits more widely to more social economic groups in order to forestall the problems of 'digital divide' (p.240).

In an effort to improve on globalization efforts in areas where technology and ethical considerations are key, more attention must be invested in developing theoretical frameworks to help explain key drivers, processes, players, and consequences of globalization. In addition new ethical policies and guidelines must be developed to ensure that that large scale globalization efforts do not lead to greater harm than good. Given the lack of trust and skepticism towards external ICT development efforts in South Africa, a policy governing responsible business practices is one interesting option. An example of this is the United Nations (2003) Global Compact Initiative, which encourages responsible corporate citizenship to help address the challenges of globalization. It is based on the assumption that economic development and poverty reduction depend on prosperity gained through efficient and responsible business practices. It is a global pact with business membership around the world with the following principles:

Human Rights

- **Principle 1:** Businesses should support and respect the protection of internationally proclaimed human rights
- **Principle 2:** Make sure that they are not complicit in human rights abuses

Labor Standards

- **Principle 3:** Businesses should uphold the freedom of association and the effective recognition of the right to collective bargaining
- **Principle 4:** The elimination of all forms of forced and compulsory labor
- **Principle 5:** The effective abolition of child labor
- **Principle 6:** The elimination of discrimination in respect of employment and occupation
- **Principle 7:** Businesses should support a precautionary approach to environmental challenges
- **Principle 8:** Undertake initiatives to promote greater environmental responsibility
- **Principle 9:** Encourage the development and diffusion of environmentally friendly technologies

Anti-Corruption

- **Principle 10:** Businesses should work against all forms of corruption, including extortion and bribery. (Global Compact, 2009)

This policy does provide a useful guide for global business ethics and has thousands of participating business around the world. On the one hand, it may prove useful since it has no real power behind it to police businesses and enforce penalties from businesses that do not wish to be members. On the other hand, it could be quite useful in the context of international development in creating a list of trusted business with good reputations that developing countries may take into consideration. However, the UN must be careful not to use such a list to coerce developing countries to partner with Global Compact members as this could be seen as a conflict of interest.

DISCUSSION QUESTIONS

- What concepts of globalization are most relevant to the current state of the knowledge society. Elaborate.
- Why is technology important to globalization?
- Technological advancement often leads to the undermining of already established businesses at a local level. For example, the growth of e-commerce over the last 10 years poses possible threats to local retail. Should something be done to prevent this?
- How could a international business ethics code (like Global Impact) have a greater positive influence on developing countries and their efforts to compete in the global knowledge economy?

DISCUSSION

Based on the above discussion, it can be seen that globalizing technology within a technoethical framework focuses on the multidimensional nature of globalization constituted by interconnected relations influencing South African society and its response to globalization processes. As was demonstrated in the case of South African ICT development how technoethical considerations are intertwined with historical, political, and social factors that have impeded progress.

REFERENCES

Albrow, M. (1996). *The global age.* Cambridge, UK: Polity Press.

Baskaran, A., Muchie, M., & Maharajh, R. (2006). Innovation system for ICT: The case of South Africa. In A. Baskaran & M. Muchie (eds.), *Bridging the digital divide: Innovation systems for ICT in Brazil, China, India, Thailand, and Southern Africa (pp. 181-215).* London: Adonis & Abbey.

Dunning, J., & Narula, R. (1996). Developing countries versus multinationals in a globalizing world: the dangers of falling behind. [MERIT, Maastricht.]. *Research Memorandum, 2,* 6–22.

Findlay, M. (2000). *The globalization of crime: Understanding transnational relationships in context.* Cambridge, UK: Cambridge University Press.

Friedman, T. (1999). *The lexus and the olive tree: Understanding globalization.* Cambridge, UK: Polity Press.

Giddens, A. (1991). *The consequences of modernity.* Cambridge, UK: Polity Press.

Global Compact. (2009). *Global Compact Initiative.* United Nations.

Held, D., & McGrew, A. (Eds.). (2003). *The Global Transformations Reader: An Introduction to the Globalization Debate.* Cambridge, UK: Polity Press.

Jeffrey (2001). Information technology, cumulative causation and patterns of globalization in the third world. *Review of International Political Economy, 8*(1), 147–162.

Mouton, I. (2003). South African Science in Transition. *Science, Technology & Society, 8*(2), 235–260. doi:10.1177/097172180300800205

Muchie, M., & Baskaran, A. (2006). General conclusions: Innovation system for ICT. In A. Baskaran & M. Muchie (eds.), *Bridging the digital divide: Innovation systems for ICT in Brazil, China, India, Thailand, and Southern Africa,* (pp. 237-243). London: Adonis & Abbey.

Robertson, R. (1992). *Globalization: Social theory and global culture.* London: Sage.

UNCTAD. (1996). *World Investment Report.* New York: United Nations.

United Nations. (2003). *The Global Compact: Corporate citizenship in the world economy.* New York: Global Compact Office.

Waters, M. (1995). *Globalization.* New York: Routledge.

ADDITIONAL READING

Burbules, N., & Torres, C. (2000). *Globalization and education: critical perspectives*. New York: Polity Press.

James, J. (2001). Information technology, cumulative causation and patterns of globalization in the third world. *Review of International Political Economy, 8,* 147–162. doi:10.1080/09692290010010281

Jhurree, V. (2005). Technology integration in education in developing countries: Guidelines to policy makers. *International Education Journal, 6*(4), 467–483.

Kozma, R., McGhee, R., Quellmalz, E., & Zalles, D. (2004). Closing the digital divide: Evaluation of the World Links program. *International Journal of Educational Development, 24,* 361–381. doi:10.1016/j.ijedudev.2003.11.014

Röling, N., Ascroft, J., & Chege, F. Y. (1976). The diffusion of innovations and the issue of equity in rural development. *Communication Research, 3,* 155–170. doi:10.1177/009365027600300204

Scholte, J. A. (2000). *Globalization: A critical introduction*. New York: St-Martin's Press.

Chapter 12
Global Technoethics and Cultural Tensions in Canada

INTRODUCTION

From the time I was a child until now I have seen changes. We once used snow shoes and now we use skidoos and trucks and to get to our land we use airplanes. The reality is that development will happen but that we would like to have a say in the pace that this happens to try to hold back and allow our people time; time for our land time to heal, and time to allow my people to adapt to these changes that are happening so fast. --Matthew Coon Come, Grand Chief of Quebec's Grand Council of the Crees.

DOI: 10.4018/978-1-60566-952-6.ch012

Winston Churchill once said that history is written by the victors. This statement from Churchill highlights the challenge that marginalized local cultures face in the global world and how important parts of their cultural history can get left behind and forgotten in the drive for national prosperity in the global economy. This chapter focuses on the cultural tensions that arise when a technology rich culture threatens the sustainability of a technology poor culture. A pilot case study of cultural tensions between aboriginal people and dominant French and English Canadian populations. This pilot study explores how technoethical considerations are intertwined with historical, political, and social factors that have threatened the sustainability of aboriginal culture in Canada. Findings suggest that more attention must be invested to ensure that that globalization efforts by technology rich dominant cultures do not lead to the demise of technology poor marginalized cultures. Given the longstanding history and broad scope of aboriginal problems in Canada efforts to revive the cultural history and identity of aboriginal people is suggested as one option to help rebuild aboriginal trust and willingness to collaborate with dominant Canadian populations on global initiatives.

BACKGROUND

What is Culture and how does it emerge? Culture can be described as the collection of symbols, language, behaviors, practices, customs, traditions, beliefs, and values held by people brought together with common needs and common challenges for survival. Culture is embedded within family and professional life, communities and villages, and organizations within society.

What is Global Culture? The idea of a global culture first became popularized by Marshall McLuhan in the early 1960s in describing the shift in society from an individualistic visual print culture to an oral culture (tribal culture) dominated by electronic media, increased interdependence, and a greater sense of collective identity. McLuhan (1962) referred to this new form of social organization he referred to as a 'global village':

Print is the extreme phase of alphabet culture that detribalizes or decollectivizes man in the first instance. Print raises the visual features of alphabet to highest intensity of definition. Thus print carries the individuating power of the phonetic alphabet much further than manuscript culture could ever do. Print is the technology of individualism. If men decided to modify this visual technology by an electric technology, individualism would also be modified (p. 158)

McLuhan and Powers' (1989)*The Global Village: Transformations in World Life and Media in the 21st Century* applied the idea of a global village to describe the challenges in contemporary life and culture due to competing cultural mindsets between visual (linear and individualistic) and acoustic (non-linear and many-centred) oriented culture.

McLuhan's technology driven global village was important because it helped raise awareness about the global culture from a mass media and marking standpoint. For example, Lash and Lury's (2007)*Global Culture Industry: The Mediation of Things.* Explore how this many-centred contemporary culture has increasingly become a commodity, giving rise to global culture industry. This work (and other similar work) argues that material objects are powerful cultural symbols which have spread through capitalism to form globally recognized brands. The notion of global village was also important in raising awareness about a new type of 'global consciousnesses created by the spread of multiple cultural symbols and the increased integration of foreign products, services, and ideas into local life and society. Robertson (1992) notes, Globalization as a concept refers both to the compression of the world and the intensification of consciousness of the world as a whole . . . both concrete global interdependence and consciousness of the global whole'(8). Taken together, global culture refers to a dimension of globalization driven by the integration of new technology and practices which promote greater cross-cultural connections and new forms of collective consciousness arising from increased diffusion of foreign cultural products, services, and ideas around the world.

Robertson's (1992)*Globalization: Social Theory and Global Culture* examines globalization and culture within the context of global and local tensions embedded within political and economic changes within an increasingly connected world. Robertson (2992) likens this new world to a global pool hall,

By endowing nations, societies, or cultures with the qualities of internally homogeneous and externally distinctive and bounded objects, we create a model of the world as a global pool hall in which the entities spin off each other like so many hard and round billiard balls. (p. 30)

Within this global pool hall, individuals live within a worldwide network of social relations are offered new opportunities to link dispersed cultures in such a way that local happenings are influenced foreign events and vice versa. But what is life like in this global pool hall? Are there enough tables for everyone to play? Who is responsible for monitoring the pool hall for proper use? Does everyone abide by the rules of the pool hall? What happens if a disagreement or fight breaks out at the pool hall?

Global culture is a highly debated area of globalization with many controversial questions yet to resolve. Does the rise of global culture and mass consumerism threaten the survival of language, as a complex symbol system difficult to transfer? Does the spread of multiculturalism and increased exposure to cultural diversity threaten to supplant local cultures? Assuming that advanced technology and greater global connectedness brings with it greater wealth and power, is there a risk that more globally oriented cultures (dominant cultures) will push out or assimilate cultures that are less globally connected (marginalized cultures)? It is with this last question that the following case addresses.

GLOBALIZING CULTURAL TENSIONS IN CANADA: THE CASE OF ABORIGINALS IN CANADA

Description

How has the presence of dominant populations in Canada (English and French Canadian Cultures) influenced aboriginal culture? How can this influenced be described (nurturing, threatening, etc.)? What structures and mechanisms (historical, political, social, economic, educational, etc) shape this influence? What social and ethical dilemmas face conflicting cultures in Canada in global economic crisis?

This case explores aboriginal culture in Canada amidst competing cultural interests of dominant French and English Canadian cultures. Recently the World Health Organization classified Canadian Indian reserves as a global disgrace for a country as rich as Canada. How it is that the state of aboriginal culture has reached the point it is at now? For case clarity, it draws mainly on events occurring involving aboriginal population in Quebec, an area of Canada with a particularly high level of aboriginal conflict. As is discussed below, aboriginal cultural tensions are systemic and influenced by underlying historical, economic, social, and political conditions, as well as key events and actions by people. Below is a breakdown of selected events linked to recent aboriginal cultural tensions in Canada.

Aboriginal people in Canada (also known as First Nations, Inuit and Métis) were the first people to inhabit Canada. They were recognized as indigenous groups in the Canadian Constitution Act of 1982 and make up approximately 4% of the Canadian population. As stated in the Constitution Act:

1. The existing aboriginal and treaty rights of the aboriginal peoples of Canada are hereby recognized and affirmed.
2. In this Act, "aboriginal peoples of Canada" includes the Indian, Inuit and Métis peoples of Canada.

Table 1. Chronology of major events connected to aboriginal cultural tension

Date	Major Events
1890s – 1990	Residential and Industrial Schools were established in locations across Canada, predominantly in Western Canada for the purpose of "killing the Indian in the child." Over 150,000 children attended these residential schools up to 1973 (Assembly of First Nations, 2009)
1961 -- 1989	Oka Crisis occurred after longstanding land dispute which culminated with the Mohawk people building a barricade blocking access to the area to be developed, which sparked racial tensions, and local police retaliation
1980s	Matthew Coon Come organized and led the fight against the massive $13 billion James Bay hydroelectric project (to sell electricity to the U.S.) which threatened to flood aboriginal land in Northern Quebec and drive aboriginal people from their villages.
1982	Aboriginal people in Canada were recognized as indigenous groups in the Canadian Constitution Act of 1982. This reaffirmed and built upon previous Laws and Acts granting rights to aboriginal people in Canada.
1990	Grand Chief of the Manitoba Chiefs Phil Fontaine (Phil Fontaine) becomes first Indian leader to go public with his story of his own abuse in residential school. This starts a trend which spreads across Canada
1990s	There is a major trend in public disclose and filing of legal lawsuits by aboriginal people for recognition of their abuse in residential schools, compensation and an apology for the inherent racism in the policy. Numerous residential school survivor groups are formed to provide support for racist treatment of residential school victims.
1991 --1996	In 1991, The Royal Commission on Aboriginal Peoples is formed and in 1996 recommends that a public inquiry be held to investigate and document residential school abuse.
Late 1990s	A number of legal settlements are reached and the Assembly of First Nations negotiations with federal government to create an out-of-court settlement process for residential school abuses.
2001	Department of Indian Residential Schools Resolution Canada is established to address out of court settlements for residential school abuse victims.
2003	Canada launches an independent dispute resolution process.
2008	Prime Minister of Canada Stephen Harper delivers public apology to residential school survivors for wrongdoing which creating a national response among aboriginal groups across Canada and more public disclosure from aboriginal people about residential school abuses.
2008 - 2009	Canada follows the U.S into an economic recession and the need to negotiate with aboriginal people for access to resources on aboriginal land increases as Canada looks for ways to boost start a failing economy.

• Note: This table is only a selection of key events and not a complete history.

3. For greater certainty, in subsection (1) "treaty rights" includes rights that now exist by way of land claims agreements or may be so acquired.

4. Notwithstanding any other provision of this Act, the aboriginal and treaty rights referred to in subsection (1) are guaranteed equally to male and female persons. (Section 35)

Despite their recognized rights under the Canadian Constitution Act, aboriginal people have struggled to maintain their land, culture, and identity within the historical conditions which threatened their way of life. This is notable in the longstanding Canadian residential school system. The Canadian residential school system was a federal initiative which led to the abuse, neglect, and death of a significant portion of aboriginal people from the 1890s to 1990s. These were originally church run schools that forced aboriginals to assimilate into European-Canadian society by separating children from their aboriginal families. It has been described as cultural genocide and a means of "killing the Indian in the child" (Assembly of First Nations, 2009). In addition to forced assimilation, the operation of residential schools also led to physical and sexual abuse, overcrowding, poor sanitation, lack of medical care, and rates death rates up to 69 percent (Arnett, 2009). Following legal lawsuits from victims of residential school, such schools were disbanded in the 1990s.

Social and political upheavals between aboriginal people and dominant Canadian populations have also contributed to cultural tensions between these groups. This is exemplified in numerous public demonstrations in the province of Quebec over land disputes. For instance, The Oka Crisis in Quebec was a pivotal event in the struggle for aboriginal human rights in Canada (MacLaine & Boxendale, 1991). MacLaine and Boxendale (1991) described how a legal dispute over land claims arose between the town of Oka and the Mohawk community of Kanesatake when the town of Oka began planning an expansion to a golf course on a Mohawk burial ground. The problem began in 1961 when the Club de golf d'Oka developed a golf course on a portion of the land believed by the Mohawks to be theirs. This led to legal protest against its construction that was rejected in 1977. In 1989, the Mohawk people built a barricade blocking access to the area to be developed, which sparked racial tensions, local police retaliation (Sûreté du Québec), and the loss of life. The situation erupted and Mohawks from across Canada and the U.S. became involved. The Royal Canadian Mounted Police (RCMP) had to be brought in to deal with the situation with little success. No solution was ever reached since the province constructed an alternative highway route (part of auto route 30) and the federal government stepped in and purchased the land in dispute, thus removing it from both parties.

A second major dispute in Quebec arose in the 1980s when Matthew Coon Come led the fight against the massive $13 billion James Bay hydroelectric project. This project threatened to flood aboriginal villages in northern Quebec. Matthew Coon Come was born in Northern Quebec and became Grand Chief of the approximately 12,000 Cree of northern Quebec. Coon Come organized a canoe trip of Cree elders from James Bay down and down the Hudson River to New York City to protest U.S plans to purchase Quebec hydro-electricity which would flood reserve land and drive aboriginals from their villages . This public demonstration was partly influenced by differing social values and poor social relations between Quebecois and Aboriginal people. The cultural values and support for technological development and global commerce from the dominant French Canadian population in Quebec was at odds with the naturalistic values and practices of the aboriginal people in Northern Quebec. As stated by The Grand Chief, Matthew Coon Come (1996):

The greatest accomplishment of my people is not that we build great dams, great pyramids, or monuments to ourselves, but to leave the land the way it is. That is the way our ancestors saw it and the way my father and grandfather saw it. That is the greatest accomplishment (Coon Come, 1996).

These two case examples help illustrate the serious historical, social, and political influences on aboriginal culture and its pervasive tension with dominant French and English culture in Canada. The underlying conditions set the stage for a number of intervening conditions and government strategies to address the events in question. In response to tragic events (residential schools, Oka Crisis), the Royal Commission on Aboriginal Peoples created in the 1990s by the Government of Canada to review past government policies affecting Aboriginal people and make recommendations to the government to improve the situation. This has led to legal settlements being awarded and official apologies from the Prime Minister in reconciliation efforts.

In terms of intervening conditions, delays in dealing with aboriginal disputes and abuses has acted as a catalyst for growing cultural tensions and loss of personal dignity and self-respect among aboriginals negatively impacted by residential schools and land disputes. Although residential schools were opposed by aboriginal people at the time of their establishment, it required over 100 years for the government of Canada to take responsibility for wrongdoing and dismantle the residential school system. Land claim disputes have also extended for decades without resolution (I.e. Oka Crisis). In other words, the recent history of abuse and delays in accountability (i.e., residential schools in Canada, legal settlements, land rights claims) appear to have further deepened cultural tensions. Another factor that must be noted is the role of the vast geographic distance between much of the aboriginal land and settlements in the northern parts of Canada from the dominant French and English populations concentrated in the southern parts of Canada. This geographical factor may have contributed to delayed public responses and limited communications between cultural groups. Without strong communication capacities and understanding of aboriginal ways of life, the challenge of resolving disputes and reaching agreements is more difficult.

The situation is now changing and technology is providing new opportunities for aboriginal people to respond to past events that created cultural tensions and to take a more active role in communicating knowledge and views more quickly than possible before the advent of telecommunications. First, social networking and new media tools give marginalized groups new ways to create multimedia productions to document and share cultural experiences and problems. For instance, a recent You Tube video (Anonymous, 2009) uses actual footage and interviews with aboriginal people involved in the Kahnawake incident where French Canadian protesters blocked a road and threw stones at cars driven by aboriginals while police stood by watching. Increased access and ease of ICT use in recent years has permitted aboriginal people to close some of the communication gap and raise public awareness about acts of aggression and violence when they occur. The access to more information more quickly could help reduce the time needed for authorities to react when cultural tensions arise. Second, mass media broadcasts and web bases news services are increasing the breadth and depth of news coverage on events affecting aboriginal people in Canada. For instance, the public apology made by Prime Minister Stephen Harper was broadcast around the world along with feedback and commentaries from aboriginal leaders.

NEW DIRECTIONS

Canada is not alone in the clash between technology rich and poor cultures at home. Other countries like Australia have gone through similar challenges. What is interesting to note is that strategic use of technology demonstrates the potential for local cultures become embedded within a global landscape, thus creating greater understanding and more opportunities to participate in decision making which affects local relations within a 'global village' (McLuhan, 1962).

One way to sustain local culture within a global context is through technology driven cultural memory building initiatives. In 2007, Boulou Ebanda de B'béri recieved a major SSHRC grant (Historical Amnesia: The Promised Land) to conduct a cultural and historical investigation of 19th-century black pioneers' in Chatham and Dawn settlements. The focus of the initiative was the role this group played in ending slavery and fighting for civil rights in Canada and the United States (University of Ottawa, 2007). According to the website report from the University of Ottawa (2007):

These settlements, often known as the "Promised Land" communities, were home to a significant group of people of African descent in the 1800s and are commonly linked to popularized stories about the underground railroad escape from slavery in the U.S. to Canada; which is only a part of a much larger story. The "Promised Land" communities generated powerful ideologies of freedom, identity, and citizen-

ship. From this ideological crucible, black Canadian women and men in the 19th century worked to abolish slavery in the United States, and to protect civil rights in Canada. Though the communities themselves were small, their influence stretched across Canada and to the farthest reaches of the Atlantic world. They were the vital center of a culture of justice that drew interracial support and forged links of freedom across the United States and Britain (University of Ottawa, 2007).

This type of cultural exploration is helping to reclaim an important part of Canada's black history (local culture) and multicultural heritage at risk of being lost in the globalizing cultural context. By collecting stories and public digital repository creation efforts, this type of initaive can provide public access to cultural artifacts, advance educational programs about black culture in Canada, and create community projects in the arts and public history. This type of work provides a model for local culture buiding (and rebuilding) within a global context. The digitalized stories can be told and retold locally and shared with other members of this local culture that have been scattered worldwide within the global world. The communication links created can continue to invigorate local culture efforts while promoting the sharing of this culture within a global context (global culture).

DISCUSSION

This discussion focused on cultural tensions of aboriginal people in Canada. It demonstrated how cultural tensions arose and how technology could be used to improve communications and promote greater understanding and sharing of perspectives needed to begin rebuilding trust and engaging in major initiative involving aboriginal people and other Canadian populations. Given that Canada and the U.S. are in an economic recession and the need to negotiate with aboriginal people for access to resources on aboriginal land, it is of vital importance for the sustainability of Canada to improve relations and collaborative effort with aboriginal people. This need is expected to further intensify as Canada looks for ways to boost start a failing economy when a major portion of natural resources in Canada are on aboriginal land. As stated by Coon Come (1996) stated, "We as aboriginal people are caught in disputes between the English and French over land which was not theirs in the first place". Therefore, Canada must find ways to rebuild relations and allow aboriginal people to preserve their culture and ways of life. This will not be easy since the approach of aboriginal culture to technology is not the same as dominant English and French Canadian populations.

DISCUSSION QUESTIONS

- How has the presence of dominant populations in Canada (English and French Canadian Cultures) influenced aboriginal culture?
- How can this influenced be described (nurturing, threatening, other)?
- What structures and mechanisms (historical, political, social, economic, educational, etc) shape this influence?
- What social and ethical dilemmas face conflicting cultures in Canada in global economic crisis?
- What barriers need to be overcome and what links need to be created to sustain local culture efforts of aboriginals while promoting the sharing of this culture within a global context (global culture)?

REFERENCES

Anonymous. (2009). *French tried to stone & kill Indians in Canada*. Retrieved February 1, 2009 from http://www.youtube.com/watch?v=ZYB-aF_MlB0

Arnett, K. (2009). *Hidden from history: The Canadian holocaust—the untold story of the genocide of aboriginal peoples*. Retrieved February 1, 2009, from http://www.hiddenfromhistory.org/

Assembly of First Nations. (2009). *Residential schools: A chronology*. Retrieved February 1, 2009 from, http://www.afn.ca/article.asp?id=2586

Coon Come, M. (1996). *Cree freedom. London: Journeyman Pictures*. Retreived January 1, 2009 from http://www.youtube.com/watch?v=7GdegbTQGA0.

Harper, S. (June 11, 2008). Indian residential school- Apology PM Stephen Harper. Retrieved February 1, 2009 from http://www.youtube.com/watch?v=wyxJ-zpYDkE

Lash, S., & Lury, C. (2007). *Global culture industry: The mediation of things*. New York: Polity.

MacLaine, C., & Boxendale, M. (1991). *This land is our land: The Mohawk revolt at Oka*. Quebec: Optimum Publishing International Inc.

McLuhan, M. (1962). *The gutenberg galaxy*. Toronto: McGraw Hill.

McLuhan, M., & Powers, B. (1989). *The global village: Transformations in world life and media in the 21st century*. Oxford, UK: Oxford University Press.

Robertson, R. (1992). *Globalization: Social theory and global culture.* Thousand Oaks, CA: SAGE.

UNESCO. (2006). *Gateway to Global Culture site launched.* Retreived February 1, 2009, from, http://portal.unesco.org/ci/en/ev.php-URL_ID=22616&URL_DO=DO_TOPIC&URL_SECTION=201.html

University of Ottawa. (2007). *University of Ottawa researcher awarded one million dollars to study historical amnesia and black Canadian settlements.* Retrieved February 1, 2009 from http://www.research.uottawa.ca/news-details_1375.html.

ADDITIONAL READING

Grübler, A. (2003). *Technology and global change.* New York: Cambridge University Press.

Lumsden, I. (1970). *Close the 49th parallel etc: The americanization of Canada.* Toronto and Buffalo: University of Toronto Press.

McGrew, A., & Held, D. (Eds.). (2003). *The global transformations reader: An introduction to the globalization debate. Cambridge.* UK: Polity Press.

Milloy, J. (1999). *A national crime: The Canadian government and the residential school system, 1879 to 1986.* Toronto: University of Toronto Press.

Modelski, G. (1972). Principles of world politics. New York: Free Press.

Murphy, R. (July, 2004). Matthew Coon Come interview. CBC News Online, July 2,2004. Retreived February 1, 2009, from http://www.cbc.ca/news/background/aboriginals/cooncome_interview.html

Taras, D. (2001). Surviving the wave: Canadian identity in the era of digital globalization. In D. Taras & B. Rasporich (eds.), *A passion for identity: Canadian studies for the 21st century (pp. 293 – 309).* Scarborough: Nelson Thompson Learning.

Section 5
Future Trends:
Shaping the Future of Technoethics

Chapter 13
Trends in Technoethics:
The Case of E–Technology A/Buse at Work

INTRODUCTION

Technology... is a queer thing. It brings you great gifts with one hand, and it stabs you in the back with the other - C.P. Snow

The widespread integration of electronic technologies in contemporary organizations is transforming organizational operations and how people work. This can partly be attributed to the advent of reliable technological infrastructures, increased workflow efficiency, and operational cost savings that can be achieved by utilizing electronic technologies. New electronic technologies allow organizational members to exchange more information in less time and with greater flexibility than possible through

DOI: 10.4018/978-1-60566-952-6.ch013

traditional means. Electronic technology common to organizations includes, but is not limited to, computer monitoring and filtering systems, surveillance cameras, IM and other chat tools, electronic mail and voicemail, Internet, audio and video conferencing tools, personal data assistants, and mobile devices. Non-work-related use of electronic technologies may include personal use of electronic technologies as well as institutional uses of counter measure electronic technologies to control personal use.

In spite of their ubiquity and remarkable benefits, there are substantial obstacles to ethically responsible electronic technology use that has given rise to debate within organizations. Of particular concern to organizations are electronic technology mis-uses with ethical and legal implications which could compromise organizations and their reputation in the public eye. Documented research on the misuses of electronic technologies within workplace contexts include, cyberloafing, network attacks, information theft, unauthorized transmission of confidential information, transmis-sion of offensive jokes, deceptive communication, exchanging online pornography, online flaming, and online identity deception (Lim, 2002, Redding, 1996; Wilson, 2005). In response to this perceived problem, many organizations have responded through counter measures aimed at controlling personal use of electronic technolo-gies; include the installation of surveillance cameras, the enforcement of electronic technology use policies, and the installation of computer monitoring and filtering systems. This too has raised ethical concerns about human rights, freedom, and privacy issues. This suggests a serious disconnect revolving around technology giving rise to a technoethical crisis in the workplace. This is particularly salient among students who are entering the workforce as new employees. According to a recent survey conducted by Harris Interactive(2008):

Young workers [age 18 to 34] are also the most likely to use their employers' tech-nology for personal reasons. Nearly three-in-four (72%) check their personal email accounts during work, and 77% use their work Internet personally compared to 69% of office workers overall (12).

Unfortunately, the reasons behind controversial uses of electronic technologies in the workplace is a relatively understudied area with a paucity of studies focusing on ethical considerations underlying electronic technology use in work contexts. What are employees' ethical stances on personal use of electronic technologies during work hours? What are employees' ethical stances on the organizational use of counter measures (e.g., surveillance cameras, restriction policies, computer monitoring and filtering software, etc) to control the personal use of electronic technologies during work hours? What would affect employees' decision to use or not use electronic technologies for personal use during work hours? In view of the importance of electronic technologies in the workplace and the seriousness of the ethical concerns connected to their use, there is a need to get an insider perspective from the user to better understand why individuals use of electronic technologies within workplace contexts. To understand the nature of this technoethical crisis, a pilot case study is presented to explore ethical perspectives on controversial electronic technology use within work settings. A university student population is selected because they are currently entering the workforce and represent the future of work life. As will be seen, findings identify reveal a disconnect between individual and organizational ethical stances that warrant further technoethical inquiry.

TECHNOETHICAL STUDY ON ELECTRONIC TECHNOLOGY

Study Description

This technoethical study explores how university students entering the workforce view personal use and countermeasures to personal use of electronic technology in work settings. Of particular interest are ethical views about electronic technologies used at work as well as the variables that affect their decision to engage in or not engage in personal use of electronic technologies during work hours. This study utilizes selected research literature and anonymous conversation data derived from student lab work completed in partial fulfillment of requirements for an undergraduate research course. Preliminary findings presented in this chapter shed light on student attitudes, subjective norms, attributions of responsibility, and other factors affecting personal use of electronic technologies usage during work hours within workplace contexts.

A grounded theory approach of Strauss and Corbin (1998) guided data collection and analysis to 'offer insight, enhance understanding, and provide a meaningful guide to action' (Strauss & Corbin, 1998, p. 12). The following research question guides this study: How do students entering the workforce view non-work related use (personal use and countermeasures) of electronic technology in the workplace? A review of the research literature on non-work related use of electronic technology within work settings and student conversations are used to base study interpretations. Conversation data is derived from online conversations guided by discussion questions focusing on ethical perspectives of electronic technology use and abuse in the workplace. Data analysis is applied to both interview data and documents. In initial coding, concepts were compared with one another and those that appeared to pertain to the same phenomenon were grouped into the main category of electronic technology abuse at work (open coding). A second tier of coding (axial coding) focused on locating links between the main categories and sub-categories

Figure 1. E-technology abuse at work

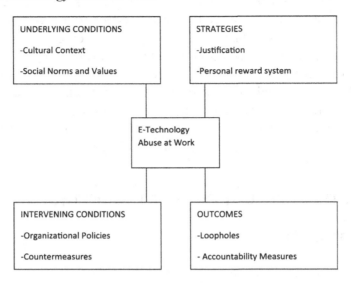

by analyzing individual categories in terms of their underlying conditions, context specific strategies, intervening conditions (interactions), and outcomes (Strauss & Corbin, 1998). The findings and interpretations are presented below.

Emergent Themes

A number of moderately related themes emerged in an analysis of selected conversation data. These were grouped into the following categories illustrated in Figure 1:

E-Technology Abuse at Work

Does E-technology abuse at work exist? This was discussed by all students with a widespread acknowledgement of its occurrence within their work place environments. Moreover, almost all students admitted to doing it themselves. One student stated that they "I definitely use the work devices for personal use and personal devices during work time—don't tell my boss!" (excerpt). Many students admitted to using their cell phones during work hours to telephone friends and family members while others spent work time on the WWW. Another student commented that she spent hours each day at work "surfing the net or playing games to pass the time" (excerpt). This follows research from Periolat (2005) stating that, "IM is being used in 83% of all enterprises in North America, either for personal non-business use or company communications." This provides a clear indication that electronic technology exists within workplace environments. The question is, what conditions give rise to this behavior and what strategies are taken in response to e-technology abuse in the work place?

Underlying Conditions

Upon close examination of conversational data, there appeared to be a moderate level of discussion about the underlying context that permitted electronic technology abuse at work to occur. Some students felt that this was a 'normal' part of contemporary work culture and they did not understand the problem of it.

In examining the theoretical literature, there appears to be cultural and social, and conditions that promote technology and allow for electronic technology abuse in the workplace "as long as my work is done and i don't have any customers in line" (excerpt). Since the pioneering writings of McLuhan, there has been a well established link between technology as an "extension of man" and entrenched within contemporary culture and society. There a visible culture of electronic technology embedded in work and life that is, as stated by Kramarae (1988), "habitual and unconscious" (p.146). This culture of technology is particularly salient with the continued growth of technological convergence and ubiquitous electronic technologies designed to be seamlessly integrated into areas of work and life. Jenkins' (2006) *Convergence Culture: Where Old and New Media* Converge provides explores the complexity and close relationship between ICT's and the role of culture. The underlying cultural context also promotes social values revolving around technology and social Rosen (2004) explored how social stereotypes link electronic technologies to the achievement of desired ideals such as wealth and power. This is believed to be due to longstanding societal views surrounding power and connectivity. Rosen (2004) states, "In the era before cell phones, popular culture served up presidents, tin-pot dictators, and crime bosses who were never far from a prominently placed row of phones, demonstrating their importance at the hub of a vast nexus," (40). Existing scholarship suggests the presence of underlying conditions which promote close relationships with technology in work and life. This does help explain why employees might abuse technology in the workplace since it is seen as connected to their cultural and social identities.

E-Technology Strategies at Work

What are the ethical stances of individuals' entering the workforce concerning personal use of electronic technologies? When discussing ethical implications of e-technology abuse at work, there was a moderate tendency to justify one's personal use of electronic technologies at work. As indicated by one student, "I don't believe that what I'm doing is causing any sort of serious harm/crime and that my actions are pretty justifiable" (excerpt). Another stated, "I send a lot of e-mails to my coworkers while I'm supposed to be working just to keep myself awake and stimulated" (excerpt). Other students justified limited use of electronic technologies for personal use as long as employee work is complete and when it does not interfere with productivity. For example, "If the devices are being used for things that are not work related and it affects the productivity of an employee I think it causes problems" (excerpt). This is consistent with a study of non-work related Internet use conducted by D'Abate (2005) found that individuals may rationalize non-work related use of the Internet during work hours in a variety of ways including, viewing themselves as capable of simultaneously conducting work and non-work related tasks, viewing non-related Internet use a justified reward for completed work, and viewing themselves as feeling tempted by the convenience of access. Zeldes et al. (2007) argue that individuals rationalize their actions in terms of overall benefits offered by the attractiveness of continual access to information:

People afflicted by Info mania are well aware they are in trouble, yet they nurture a feeling that they should stay the course and not fight back. They rationalize that the modern world revolves around communication, so they should always be connected or lose vital information. Many believe that there is no better way" (p. 3).

Intervening Conditions

What is the level of knowledge and ethical stance of individuals' entering the workforce concerning countermeasures to control e-technologies in the workplace? When asked about the use of counter measures (e.g., surveillance cameras, restriction policies, computer monitoring and filtering software, etc) to control the personal use of electronic technologies during work hours, responses were mixed. Levels of knowledge and opinions differed concerning security measures used by an organization to prevent personal use of E-technologies. While some students has no idea if countermeasures were being used, others were aware of countermeasure used but did not know the extent to which countermeasures were used or where they were used. As indicated by one student, "when we sign our contract, we sign a paper which basically says that we will not go onto specific types of websites. It even addresses that that we promise not to forward non-work related things like jokes to each other through our work email" (excerpt).

Ethical stances of individuals' entering the workforce concerning countermeasures to control e-technologies in the workplace varied. Some did not oppose the organizational use of counter measures since, "it is their equipment, money and time" (excerpt). Others believed that the organizational use of counter measures is not necessary because it invades employee privacy. It was suggested that employees need to be trusted to do their work and to uphold a high standard of professional conduct. A third group thought that compromise was needed for strategic use of countermeasure types and areas of application. One student indicated, "I think that personal use of electronic technologies during work hours is unfair, but it happens. Companies should monitor how much it happens and dole out punishments only if it's interfering with the amount of work an employee produces" (excerpt). Another student reasoned, "I don't think employers should be allowed to read e-mails or record phone conversations to make sure they are work-related" (excerpt). Another student added, "some measures are more appropriate than others. Surveillance cameras and computer monitoring are invasive methods that can have ethical repercussions" (excerpt). Existing research has not yet explored employee perspectives on the variety of countermeasure types that exist and how they should be implemented.

Consequences

What would affect individuals' decision to use or not use electronic technologies for personal use in the workplace? Although many students justified their abuse of electronic technologies in the workplace, this was conditional on expectations of no negative consequences. As stated by one student, "The only thing I guess that would affect my decision to not use electronic technologies for personal reasons during work is if there were real serious consequences provided for doing so" (excerpt). This raises additional questions concerning the existence and nature of negative consequences. According to the research literature there are no laws against organizational use of countermeasures to control e-technologies in the workplace in Canada. Levin (2007) states, "[T]he structure of Canadian statutory tort of invasion of privacy and permits employers to introduce measures of surveillance and monitoring to the workplace with impunity" (p 316). This would entail that organizations do have the right and power to imposed consequences but what should they be? While some students speculated on organizational policies that would deal with these matters, other students felt it should be dealt with on a case to case basis. As indicated by one student, "Let people have the resources. If they abuse them constantly, then either remove said resources or fire them. It's best to treat this issue on a case-by-case basis rather than spoiling it for everyone."(excerpt).

DISCUSSION

This research explored the abuse of electronic technologies in the workplace and use of counter measure electronic technologies within organizations. Findings highlighted how many students entering the workforce openly admitted to electronic technology abuse at work and did not believe there to be an ethical problem unless it interferes with work productivity. This view was supported by research scholarship on increasing technological convergence within contemporary culture and the presence of social values which link power and prestige to technology connectedness and constant information access. These cultural and social beliefs are in conflict with organization rules and regulations which prohibit personal electronic technology use in the workplace during work hours. Further research should explore this social and ethical disconnect between organizations and those that work within them. It was also found that students had mixed ethical stances concerning the use of counter measures to block personal use of electronic technology in the workplace. While some students believed that such measures were an organizational right, others believed that it undermined trust and violated individual privacy rights. This suggests a schism in ethical stances between individuals and organizations concerning non-work related electronic technology use in the workplace.

Moreover, this study reveals a gap the research literature on employee monitoring strategies in the workplace. While some students were aware of such measures, many others were unsure whether they were being used and how they were being used in their workplace. Based on this study, it appears that the visibility of information for employees on workplace monitoring is an issue. It was also found that employee acceptance or rejection of such employee monitoring strategies in the workplace is complicated and depends on the type as well as how it is applied. While a small body of research has examined organizational implications of surveillance cameras (Whittaker, 1995; Zweig & Webster, 2002), electronic technology use policies (Martin, 1999), and computer monitoring and filtering systems (Alge, 2001; Urbaczewski & Jessup, 2002), research has not yet explored employee perspectives and ethical stances on the variety of countermeasure types that exist. Nor has research explored the communication strategies of organizations to make information on workplace monitoring available to its employees. Grunewald, Halpern, and Reville (2008) argue that "…if employees understand that such a policy is not being instituted to harm them, or because the management does not trust the employees, but instead to benefit all the company's stakeholders, then this may help serve to prevent any potential negative effects" (Grunewald et. al., 2008, p.178).

Because electronic technologies are becoming more deeply intertwined in work and life, and because this can cause these boundaries to become increasing blurred (Saetre & Sornes, 2006), it is important for the public to understand the role that electronic technologies play in helping individuals decide their work ethic in navigating this challenging work-life balance. Given the increasing sophistication and decreased size of electronic technology devices available, concealing personal technology use in the workplace is continually challenging organizational efforts to monitor. In light of this, this research suggests more work needs to be done to leverage understanding of employee and organizational values to organito reach a consensual ethical stance that all stakeholders will follow. This is validated by current views on increasing employee participation in organizational policy development for appropriate technology use in the workplace. Grunewald, Halpern, and Reville (2008) state "The policy should also be instituted only after public notice to all concerned" (p.179). This suggests that employee monitoring strategies to regulate personal electronic technology use may be more widely accepted if employees are including in key decisions about how it is integrated into the workplace. More work is needed on participatory models which promote mutual decision-making to help leverage effective employee monitoring strategies that employees and employers will agree to. Another option would be to create a code of ethics to help guide organizations in integrating employee monitoring strategies to regulate personal electronic technology use. Marx and Sherizen (1986) recommended that "[companies] should have a code of ethics regarding monitoring [, since privacy] is best protected when monitoring is minimally intrusive, directly relevant to job performance, and visible." (p.63). Determining the best fit between employee and employee technoethical relations governing non-work related technology use at work is a complex issue with much more work to be done.

DISCUSSION QUESTIONS

- What would a code of ethics on employee monitoring entail and who should create it? (e.g., employees, organization, government, international body, etc.).
- What should the punishment be for a violating a code of ethics on employee monitoring and who should be responsible for enforcing it?
- What is your ethical stance on using technology for personal use in the work place?
- What counter measure should taken by your organization to diminish personal use of technology are you aware?

REFERENCES

Alge, B. J. (2001). Effects of computer surveillance on perceptions of privacy and procedural justice. *The Journal of Applied Psychology, 86*(4), 797–804. doi:10.1037/0021-9010.86.4.797

D'Abate, C. P. (2005). Working hard or hardly working: A case study of individuals engaging in personal business on the job. *Human Relations, 58,* 1009–1032. doi:10.1177/0018726705058501

Grunewald, D., Halpern, D., & Reville, P. (2008). Management and legal issues regarding electronic surveillance of employees in the workplace. *Journal of Business Ethics, 80*(2), 175–180. doi:10.1007/s10551-007-9449-6

Harris Interactive. (2008). *Journal Broadcast Group Survey,* (pp1-15). New York: Journal Broadcast Inc.

Jenkins, H. (2006). *Convergence culture: Where old and new media converge.* New York: NYU Press.

Kramarae, C. (1988). *Technology and women's voices: Keeping in touch.* New York: Routledge.

Lavoie, J. A., & Pychyl, T. A. (2001). Cyberslacking and the procrastination superhighway: A web-based survey of online procrastination, attitudes, and emotion. *Social Science Computer Review, 19*(4), 431–444. doi:10.1177/089443930101900403

Levin, A. (2007, September). Is workplace surveillance legal in Canada? (Report). *International Journal of Information Security, 6,* 313–321. doi:10.1007/s10207-007-0026-x

Lim, V. K. (2002). The IT way of loafing on the job: Cyberloafing, neutralizing and distraction versus destruction. *Cyberpsychology & Behavior, 9*(6), 730–741.

Martin, J. (1999, March). Internet policy: Employee rights and wrongs. *HRFocus, 75,* 11.

Periolat, J. (2005, March). The neglected intrusion. *Communique Newsletter, 42*(3), 18–21.

Redding, W. C. (1996). Ethics and the study of organizational communication: When will we wake up? In J. Pritchard (Ed.), *Responsible communication: Ethical issues in business, industry, and the professions (pp. 17-40).* Cresskill, NJ: Hampton Press.

Rosen, C. (2004). Our cell phones, ourselves. *New Atlantis (Washington, D.C.), 6*(Summer), 26–45.

Saetre, A., & Sornes, J. (2006). Working at home and playing at work: Using ICTs to break down the barriers between home and work. In S. May, (Ed.), *Case studies in organizational communication: Ethical perspectives and practices,* (pp. 75-85). Thousand Oaks, CA: Sage.

Strauss, A., & Corbin, J. (1998). *Basics of Qualitative research: Grounded theory procedures and techniques.* London: Sage.

Urbaczewski, A., & Jessup, L. M. (2002). Does electronic monitoring of employee Internet usage work? *Communications of the ACM, 45*(1), 80–83. doi:10.1145/502269.502303

Whittaker, S. (1995). Rethinking video as a technology for interpersonal communications: Theory and design implications. *International Journal of Human-Computer Studies, 42*(5), 501–529. doi:10.1006/ijhc.1995.1022

Wilson, C. (2005). Computer attack and cyberterrorism: Vulnerabilities and policy issues for Congress.

Zeldes, N., Sward, D., & Louchheim, S. (2007). Infomania: Why we can't afford to ignore it any longer. *First Monday Peer-Reviewed Journal on the Internet, 12*(8). Retrieved June 1, 2008, from http://firstmonday.org/htbin/cgiwrap/bin/ojs/index.php/fm/article/view/1973/1848

Zweig, D., & Webster, J. (2002). Where is the line between benign and invasive? An examination of psychological barriers to the acceptance of awareness monitoring systems. *Journal of Organizational Behavior, 23*(5), 605–633. doi:10.1002/job.157

ADDITIONAL READING

Anandarajan, M., & Simmers, C. (Eds.). (2004). *Personal web usage in the workplace: A guide to effective human resources management.* Hershey: Idea Group Inc.

Block, W. (2001). Cyberslacking, business ethics and managerial economics. *Journal of Business Ethics, 33,* 225–231. doi:10.1023/A:1012002902693

Garrett, K., & Danziger, J. (2008). Disaffection or expected outcomes: Understanding personal Internet use during work. *Journal of Computer-Mediated Communication, 13,* 937–958. doi:10.1111/j.1083-6101.2008.00425.x

Gattiker, U., & Kelley, H. (1999). Morality and computers: Attitudes and differences in moral judgments. *Information Systems Research, 10*(3), 233–254. doi:10.1287/isre.10.3.233

Griffiths, M. (2003). Internet abuse in the workplace: Issues and concerns for employers and employment counselors. *Journal of Employment Counseling, 40*(2), 87–96.

Griffiths, M. (2003). Internet abuse in the workplace: Issues and concerns for employers and employment counselors. *Journal of Employment Counseling, 40*(2), 87–96.

Harrington, S. (1996). The effect of codes of ethics and personal denial of responsibility on computer abuse judgments and intentions. *MIS Quarterly, 20*(3), 257–278. doi:10.2307/249656

Levitt, H. (November 5, 2008). Surfing porn can still get you fired. *Financial Post*. Retrieved November 13, 2008 from http://www.financialpost.com/working/story.html?id=933695

Marshall, K. (1999). Has technology introduced new ethical problems? *Journal of Business Ethics, 19*(1), 81–90. doi:10.1023/A:1006154023743

Marx, G., & Sherizen, S. (1986). Monitoring on the job: How to protect privacy as well as property. *Technology Review, 89*(8), 63.

Mastrangelo, P. M., Everton, W., & Jolton, J. A. (2006). Personal use of work computers: organizational justice. *Journal of Organizational Behavior, 23*, 675–694.

Chapter 14
The Future of Technoethics

INTRODUCTION

As the title of this book suggests, the knowledge society is in a state of change and transition, rather than a being a fixed point in time and history. One of the greatest objectives (and challenges) in the 21st century is to complete the transition into a knowledge society within the ever expanding landscape of information, technological advancement, and ethical tensions. The western world has survived the fallout of the 'information bomb" and the "collapse of globalization". Society has struggled through information overload, extended work weeks, increased time and space compression, loss of personal meaning, and a variety of new inequalities created through eroding social institutions and fragmenting social relations at a global level. Throughout this struggle, technology has acted as a focal point and catalyst, spurring ethical debates and widespread concern over the direction of society and human life.

DOI: 10.4018/978-1-60566-952-6.ch014

Technoethics is a key component in the advancement of the evolving knowledge society because of the central importance of both technology and humans. As demonstrated in this book, technoethical inquiry provides important insights and direction to help refocus attention on key areas of technology related human activity that raise widespread ethical concern and debate which necessitate special attention and care.

THE EVOLVING RELATIONAL SYSTEM OF TECHNOLOGY AND SOCIETY

Finding solutions to many new ethical issues revolving around technology in society are not obvious with approaches developed thus far. In response, this book advances the study of technology and ethical issues arising within a complex system that is constantly evolving and changing within society. This systems view places technology at the centre of human activity and human activity at the centre of technological progress and innovation. Because the relationships held with technology are always changing in multiple ways and at multiple levels within society, a relational system is ideally suited to discern the subtle ebb and flow of these changing relationships. Key questions that come to may include: What meaning does technology have in a complex relational system of humans and society? What are the positive and negative aspects of technology advancement for individuals and society? To what extent is technology an instrument of humans and to extant extent are humans and instrument of technology? How do individuals and organizations balance human and technological needs? What ethical aspects of technology can be derived by developing a system approach?

The rapid acceleration of knowledge production in technological contexts increases the need for in-depth mastery of techneothics for leveraging organizations. This compounds the already challenging task of accessing and sharing tacit knowledge within individuals and organizations that is core to successfully harnessing knowledge production, particularly in scientific and technological areas which are complex, highly specialized, and difficult to communicate. Part of the problem is that scientific and technological knowledge creation and management is being mismanaged, or to put it another way, knowledge management is currently dominated by professionals who are not knowledge experts in the domains they are trying to manage. A significant number of knowledge managers come from Business with MBA degrees and have no knowledge (except domain specific business knowledge) in the domain of technology and ethics. Accessing technoethical knowledge in the domain of specialist areas (e.g., nanotechnology, biotechnology, ICT) requires expertise to help translate technical information for the organization and other stakeholders that need it. An MBA does not provide such domain specific knowledge, which is why knowledge management initiatives often fail. These knowledge managers are outsiders who do not know enough about what they are managing to properly manage. Knowledge management must be managed from the inside and this requires real domain specific expertise in ethical inquiry and technological expertise that must emerge from within the specialty area itself. Given the rapid expansion of specialized knowledge in a technological context, there will be continued pressure for future knowledge management to be managed from within by technoethical domain experts well equipped to leverage the tacit and explicit dimension of knowledge and expertise for leveraging organizations. This should help further legitimize ethical scientific and technological innovation within the knowledge based economy of tomorrow.

The other challenge, which is more systemic and difficult to remedy, is the problem of getting experts to share knowledge. In the current context, many workers resist sharing their knowledge within the organizations they work in. There is a somewhat justified fear in giving up professional secrets and training because it may reduce the need for highly paid professions who can be replaced by technicians who can be paid much less. "I will not share my knowledge because if I do, the organization may take my system and replace me with someone cheaper who will follow the system I created." This fear is supported by the current climate of organizational life marked by lower employment stability and more organizational changes than typically occurred 50 years ago. This knowledge sharing problem is not helped by the presence of knowledge managers who are outsiders that lack subject matter expertise needed to appreciate the professionals from which the knowledge flows. It is in the hopes that a shift to community based models of decision making and improved knowledge creation and management approaches concerning ethical aspects of technology will help solidify the bonds of trust and respect that will ease the fear and tension created by giving unscientific and technological knowledge.

FUTURE TRENDS

Given the increasing integration of new technologies in many areas of human activity, the future of Technoethics is promising, but also challenging. Since ethical considerations are embedded in diverse and rapidly changing domains of technology, it is a challenge for scholars to stay abreast of key ethical issues when they arise. This is compounded by the fact that Technoethics is an interdisciplinary field drawing from niche areas that cut across many disciplines and fields. It is expected that these niche areas will continue to expand while new branches of Technoethics emerge as technology progresses in the 21st century. For this reason, it is important for technoethical scholarship to discover new strategies for advancing work conducting research in Technoethics. One useful rule of thumb suggested in Tavani (2007) involves the labelling of controversial practices as ethical problems accompanied by an analysis and synthesis of factual data associated with these problems. A more rigorous approach, and the one advocated in this book, highlights the importance of using a disciplined approach to technoethical inquiry to help inform the development of new models for guiding technology related decision making. As a biotechnocentric oriented field, the future viability of Technoethics depends partly on the ability of talented scholars to uncover social and ethical relations between technology and living entities. The *Handbook of Research on Technoethics* acknowledged the need to advance general technoethical policies and principles to accommodate the rise of unique ethical challenges made possible through scientific and technological advancement. To this end, Luppicini (2008) posited the Law of Technoethics to help situate technoethical inquiry within human activity affected by technology. The Law of Technoethics holds that ethical rights and responsibilities assigned to technology and its creators increases as technological innovations increase their social impact. If properly applied, such a law could address the need for professional responsibilities among professions who create technology. This follows Bunge's (1977) prescriptions in the original conceptualization of Technoethics and is consistent with efforts to devise principles for guiding work conducted within individual branches of Technoethics like Computer Ethics (Moor, 2005). Work is needed to help align core branches of Technoethics with the Law of Technoethics to ensure that the increased number and diversity of ethical relations match corresponding increases in the development of new technology under various conditions and within different contexts.

CONCLUSION

This is where this book ends and where further work on Technoethics must begin. Throughout this book, an interdisciplinary body of scholarship has been pulled together to demystify technoethical inquiry and create the groundwork for continued research in evolving areas ethics and technology. In this way, Technoethics acts as a converging lens using technology and ethics as the focal point. It brings together a diverse array of scholarship that sheds light on the most challenging problems in society today while attempting to produce a meaningful image. As with any scholarly synthesis of this scope, the resulting image created needs additional focusing. There may be spottiness and blurring in parts of the image created that need to be further refined and focused to meet the changing lighting conditions within our rapidly expanding technological landscape.

The most pressing need of our times is to nurture the ethical use of technology in all areas of work and life. This requires a consolidated effort in society to remove technologies which limit opportunities for meaningful participation in social and political life. Within this advancing technology mediated landscape, we have to promote the use of technology that meets the approval of all those affected in order to situate ourselves and others in what is mediated or risk losing ourselves to the technological system of control that we created. The question is simple: What relationship with technology best serves the needs and interests of society and how can this relation be cultivated.

REFERENCES

Luppicini, R. (2008). Introducing technoethics. In R. Luppicini & R. Adell (eds.), *Handbook of research on technoethics*, (pp. 1-18). Hershey, PA: Idea Group Publishing.

Moor, J. H. (2005). Why we need better Ethics for emerging technologies. *Ethics and Information Technology*, 7, 111–119. doi:10.1007/s10676-006-0008-0

Tavani, H. T. (2007). *Ethics and technology: Ethical issues in an age of information and communication technology*. Hoboken, NJ: John Wiley & Sons.

Appendix:
Interviews with Experts in the Field of Technoethics

INTERVIEW WITH TECHNOETHICS EXPERT, MARTIN RYDER, SUN MICROSYSTEMS

Question: How did you get involved in research in Technoethics and why?

I was a technophile at an early age. I loved experimenting with electricity as a child and I started college in pursuit of an engineering career. I came of age in the 1960s when the United States was embroiled in an ill-conceived, protracted war against a small Third-World nation. The innocent views I held about technology in my early years had been shocked into rude arousal by the dubious aims of that war, aroused further as I engaged with peers in political and philosophical discussions about technology and society, and aroused even further as I immersed myself in dystopian narratives of popular writers like George Orwell (1949), Herbert Marcuse (1964), Jacques Ellul (1964) and others.

Ellul put forward a biting, incisive analysis of our technological civilization, where the role of modern technology was reframed from that of a servant of humankind to that of a controlling overlord. Ellul pictured a bleak world controlled by a technical rationality that threatens our humanity and which lays waste to our natural habitat. Marcuse spoke of a technocracy that strips human beings of their uniqueness and subjectivity, a technocratic ethic that fosters conformist, passive, and homogenous behaviour, and a 'one dimensional' manipulative force that uses mass media to reshape active citizens into passive, mindless consumers. Both Ellul and Marcuse offered a vivid picture of a technical world not unlike the fictional world represented in Orwell's classic.

I emerged from the '60s still a technophile, but counting myself within the 'counterculture' (Reich, 1970). I embraced a strong critique of technocracy in contexts where technology replaces human agency in shaping our world and ourselves.

In the early 1990s when the World Wide Web was just emerging, I was an early adopter. I constructed the first academic web at my University. I observed those who began to embrace this technology, and I was keenly interested in those who did not. I put up content on the topic of 'Luddism and the Neo-Luddite Reaction' and here I posted articles about the original Luddites of 1811 along with the views of contemporary critics of modern technology. Amidst the frenzied rush into the so-called 'information age', I felt it was important to stay in touch with those who harbored a healthy distrust of the changes taking place and hearing the ethical arguments they put forward. Eventually I developed another page on the broader topic, "Theory of Technology" where I posted links to articles surrounding multiple phases of the philosophy of technology, including technoethics.

A Web page is a gregarious object with an amazing ability to attract people with common interests. The work I published on the Web attracted anonymous peers, collegial communities, and it led to multiple opportunities for collaboration with distinguished scholars whose work I have long admired. It is in this context that I found myself actively engaged in research and writing on the ethical aspects of technology.

Question: What are your main research interests in Technoethics and what findings in your work have you found most fascinating thus far?

Humans assert their intentions through their artifacts. Since ethics has to do with intentionality, it is no leap to suggest that the field of technoethics is as expansive as the field of ethics itself. The work I have done under the banner of technoethics is scattered, and I prefer to think of it as 'investigative journalism' rather than pure research. Most of my published contributions have been solicited. I have written about Semiotics in contexts of communication ethics and mass media. I have raised questions about the propriety of proprietorship with respect to patents and copyrights. I have written about the so-called 'Digital Divide' and about recent attempts to address information poverty. I am interested in the social construction of technology and in technology adoption from the standpoint of actor-network theory (Callon, Latour, Law et. al.).

There are a few topics I would like to address that demand greater urgency. Among these is the contemporary manifestation of a problem raised by Aristotle in 350 B.C.E.. In his Nicomachean Ethics (VII.1-10), Aristotle introduced the term 'incontinence' (akrasia) to describe a condition of weakness or lack of mastery that renders a person incapable of acting in accord with their better judgement. At a time in history when global warming looms as the primary threat to the habitat of all living species, our society seems to exhibit a profound incontinence when it comes to accepting responsibility for this crisis and for making life-style changes that we know are necessary to hold back the tide of climate change. We seem to have high hopes that our children and grand children will solve these problems, so we don't have to. We remain complacent in our comfortable, energy-rich life style, preferring short-term jerks and starts such as 'drill baby drill' rather than to tackle the real hurdles that demand immediate action and committed determination to solve the long term climate crisis. To me, this is the supreme ethical issue of our time.

Question: What do you consider to be the main ethical challenges in the design and development of technological innovation?

There is certainly no lack of innovation today. We see it all around us. Each day something new comes along and it is a struggle in this modern life to keep up with such constant change.

But to your point, there are challenges that inhibit an innovation from moving forward to the development of an actual product. The hurdles are typically economic. But there are ample artificial barriers that prevent an innovation from becoming reality. Chief among these are the legal claims of patents and copyrights that grant exclusive rights of development to the inventor or his/her assignee. Few of us would deny the ethical right of an inventor to reap the benefits from the labor and creativity invested in an innovation. But in today's world of corporate enterprises, ownership of an idea is rarely held by the inventor and the decision to develop an innovation is almost never made by the person who came up with the idea.

The notion of 'intellectual property' suggests that knowledge is an object that can be owned, handled, and traded like a watermelon or a sack of potatoes. The notion of 'knowledge as object' has a long tradition in the West, stemming from the Middle Ages where a written text was considered the embodiment of knowledge. Higher education in the Middle Ages amounted to the practice of transmitting a text from a faithful reading by a lecturer to the transcriptions of pupils. In early grammar schools, knowledge was viewed as a stream of objective facts that flowed from the teacher to pupils.

Today, educators, philosophers and cognitive scientists hold a completely different view of knowledge. We understand knowledge as a social construction that is not contained in any single text, invention, object, or procedure, but in the shared interaction of individuals with other individuals and with their environment, each person bringing his or her own unique history and perspective to the collective commons. Like language, knowledge is an expression of culture. Ideas emerge out of this cultural soup and the notion that one person can claim exclusive rights to the commons seems absurd. The constructivist philosopher Rousseau (1762) considered such a practice tantamount to robbery.

The ethics behind 'intellectual property' is undergoing healthy re-examination within the engineering community today. The Free and Open Source Software movement is alive and well, and the benefits of software void of ownership restraints are beginning to be recognized even by major corporate contributors. In 2005, Sun Microsystems released its well respected Solaris operating system to the open source community. This enabled Sun to bundle Solaris with multiple open source applications including the FireFox Web browser, the PostgreSQL database, and NetSNMP system management software. More importantly, it opened the opportunity for software developers from anywhere to create applications for integration with Solaris without fear of constraints of proprietorship.

INTERVIEW WITH LORENZO MAGNANI, UNIVERSITY OF PAVIA

Question: How did you get involved in research in Technoethics and why?

During my teaching philosophy of science and ethics in the last decade in US at Georgia Tech and Cuny research in the field of distributed cognition and abductive reasoning suggested to me the idea that also morality is distributed in our technological world in a way that makes some problems particularly relevant to ethics, such as for instance in the case of ecological imbalances, the medicalization of life, and advances in biotechnology. In this perspective artifacts become what I call moral carriers and mediators. The artifact is an example of a moral mediator in the sense that it mediates – objectively, over there, in an external structure – positive or negative moral effects. Many complicated external moral mediators can also *redistribute* moral effort: they allow us to manipulate objects and information in a way that helps us overcome the paucity of internal moral options – principles and prototypes, etc. – currently available to us. I also think that moral mediators can help explain the macroscopic and growing phenomenon of global moral actions and collective responsibilities resulting from the "invisible hand" of systemic interactions among several agents at local level.

Looking at the way human beings have delegated intrinsic value to artefacts and other external entities has also suggested to me that we can grant back to humans new ethical worth. The system of designating certain animals as endangered, for example, teaches us that there is a continuous delegation of moral values to externalities; it may also cause some people to complain that wildlife receives greater moral and legal protection than, for example, disappearing cultural traditions. I wondered what reasoning process would result in a non-human thing's being valued over a living, breathing person and asked myself what might be done to elevate the status of human beings. One solution, I believe, is to re-examine the respect we have developed for particular externalities and then use those things as a vehicle to return value to people.

Question: What are your main research interests in Technoethics and what findings in your work have you found most fascinating thus far?

Knowledge, I believe, is fundamental to ethical reasoning and it must therefore be considered a duty in our morally complex technological world. If we are to regard knowledge in this new light, we must first understand how knowledge can render an entity moral. For example, attitudes toward women have greatly evolved over the centuries in western society, and as our societies have gained greater knowledge, we have ascribed new kinds of value to women. As a result, the cultural default setting is generally that women have an "intrinsic" worth equal to men's.

If acts of cognition can influence moral value, I contend that we can improve the lot of many, many people by altering the way we think about them, and one way to do so is to treat them like things. This notion, of course, flies in the face of Kant's maxim that people should not be regarded as a means to an end - that is, that they should not be seen as "things." Some things – especially technologically artifacts - are treated with greater dignity than many people; I argue, consequently, that humankind will benefit if we can ascribe to people many of the values we now associate with such highly regarded non-human entities, and to that end, I suggest a new maxim—that of "respecting people as things." In this new ethical orientation, things with great intrinsic value become *moral mediators*: as we interrogate how and why we value such things, we can begin to see how and why people can (and should) be similarly respected. In this way, these things mediate moral ideas, and in doing so they can grant us precious, otherwise unreachable ethical information that will render many of our attitudes towards other human beings obsolete. The consequent concept of "respecting people as things" provides an ethical framework allowing us to analyze many aspects of the modern human condition, like for example the medicalization of life and the effects of biotechnologies, as I will more clearly explain in my response to question.

Question: Based on work in Morality in a Technological World: Knowledge as Duty, why are traditional ethical theories not adequate to deal with current ethical problems created by new technologies?

Modern technology has precipitated the need for new knowledge; it has brought about consequences of such magnitude that old policies can no longer contain them. As Hans Jonas has observed, we cannot achieve a relevant, useful new body of ethics by limiting ourselves to traditional ethics: he points out that Immanuel Kant, for example, assumed that "there is not need of science or philosophy for knowing what man has to do in order to be honest and good, and indeed to be wise and virtuous." Kant also says "[ordinary intelligence] can [...] have as good a hope of hitting the mark as any philosopher can promise himself," and goes on to state that "I need no far-reaching ingenuity to find out what I have to do in order to possess a good will.

Inexperienced in the course of world affairs, and incapable of being prepared for all the chances that happen in it," I can still ascertain how to act in accordance with moral law. Finally, Kant observes that even if wisdom "does require science," it "in itself consists more in doing and in not doing than in knowing," and so it needs science "not in order to learn from it, but in order to win acceptance and durability for its own prescriptions." Jonas concludes: "Not every thinker in ethics, it is true, went so far in discounting the cognitive side of moral action."

Question: What is the importance of "moral knowledge" and how can it be used in how can it be used to help address technoethical problems? How can moral knowledge be leveraged from considering people as if they were things (rather than ends in themselves)?

As global stakes grow higher, knowledge becomes a more critically important *duty* than ever before and must be used to inform new ethics and new behaviors in both public policy and private conduct. This means that existing levels of intelligence and knowledge in ethics are no longer sufficient; today, ethical human behavior requires that we assume *long-range responsibility* commensurate with the scope of our power to affect the planet, power that has grown exponentially over the years. I contend that improving human dignity in our technological era requires that we enhance free will, freedom, responsibility, and ownership of our destinies. To do so, it is absolutely necessary to respect knowledge in its various forms, but there are other ideas to consider as well: first, knowledge has a pivotal role in anticipating, monitoring, and managing the hazards of technology, and second, it has an intrinsic value that must be better understood, as do general information and metaethics itself. Knowledge is duty, but who owes it to whom? And is it always a right as well as an obligation?

I clearly describe some limitations of my motto "knowledge as duty" , for example in the well-known case of cyperprivacy. We must also remember that the proliferation of information carries some risks, primarily in the areas of identity and cyberprivacy. The Internet and various databases contain an astounding volume of data about people, creating for every individual a sort of external "data shadow" that identifies us or potentially identifies us, and the possible consequences must be examined. To avoid being ostracized or stigmatized, people must be protected not only from being seen but also from "feeling" too visible, which impairs the ability to make choices freely. And when sensitive information about a person is properly shielded, he or she is much less susceptible to exploitation and oppression by those seeking to use data unscrupulously

I contend that the modern undermining of Immanuel Kant's distinction between instrumental value (based on ends and outcomes) and intrinsic value (ends in themselves) results from the blurring of traditional distinctions between humans and things (machines, for example) and between natural things and artifacts. I am convinced that this shift in thought in some sense contradicts the Kantian idea that we should not treat people as means, a notion often cited by those who are suspicious of biotechnologies. Indeed, when one condemns human cloning, one usually appeals to Kant's moral principle that a person should not be treated simply as a means to other people's ends; an example might be the biotechnological cloning of a human being solely for use as a bone-marrow donor. This kind of argumentation underlies many declarations against cloning and other biotechnology, and all of these objections are linked to the central problem of human dignity. To use people for cloning would be to trample on their autonomy; to create unneeded embryos during in-vitro fertilization (IVF) would be an instrumentalization that manipulates human beings. Even if, as Kantians fear, such technology makes it impossible to avoid treating people as means, and cloning technology does indeed have the potential to produce "monsters," all is not lost: it is certainly possible to build new ethical knowledge to manage these new entities and situations. Moreover, those people, as amalgamations of the human and the artificial, will clearly seem more "thing-like" than type of person we have encountered throughout history. Does this not heighten the urgency of learning to "respect people as things"?

Question: What do you consider to be the main moral controversies connected to technologies and the world we live in (Please give 2-3 examples) and what role does philosophical inquiry play in addressing these controversies?

First of all some central methodological aspects of practical decision making and of ethical knowledge and moral deliberation have to be clarified. What is the role of reasons in practical decision making? What is the role of the dichotomy abstract vs. concrete in ethical deliberation? What are ethical "reasons"? In my book I reconsider and re-evaluate some aspects of the tradition of "casuistry" and analyze the concept of abduction as a form of hypothetical reasoning that clarifies the process of "inferring reasons." I contend that morality is the effort to guide one's conduct by reasons, that is, to do what there are the best reasons for doing while giving equal weight to the interests of each individual who will be affected by one's conduct. I also add that: 1) it is necessary to possess good and sound reasons/principles applicable to the various moral problems, and 2) we need appropriate ways of reasoning – inferences – which permit us to optimally choose and apply the available reasons. I also illustrate that "abduction" – or reasoning to hypotheses – is central to understanding some features of the problem of "inferring reasons." I contend that ethical deliberation, as a form of practical reasoning, shares many aspects with hypothetical explanatory reasoning (selection and creation of hypothesis, inference to the best explanation) as it is described by abductive reasoning in science. Of course in the moral case we have reasons that support conclusions instead of explanations that account for data, like in epistemological settings. To support this perspective, I propose a new analysis of the "logical structure of reasons, which supports the thesis that we can look to scientific thinking and problem solving for models of practical reasoning. The distinction between "internal" and "external" reasons is fundamental: internal reasons are based on a desire or on an intention, whereas external reasons are, for instance, based on external obligations and duties we can possibly recognize as such. Some of these external reason can be grounded in technological artefacts considered as moral carriers and mediators of various types.

Thus far, I consider how new technology and changing economic circumstances have generated the need for new moral knowledge, and examine the social and cultural contexts in which moral reasoning takes place. Yet to be accomplished, however, is an analysis of moral reasoning's cognitive features and constraints, and to that end, I explore how recent research in epistemology, cognitive science, artificial intelligence, and computational philosophy offer new ways for us to understand ethical knowledge and reasoning. I believe we can find models for practical reasoning in scientific thinking and problem solving – an appropriate source given the fact that science and technology underlie many of the cultural changes that have triggered the need for a new approach to ethics. It is worth remembering here that ethical knowledge, like scientific or other kinds of knowledge, is created and used by human beings and is, therefore, fundamentally related to several cognitive and epistemological concern. First of all I show that ethical knowledge and reasoning are expressed not only with words at a verbal/propositional level but also through model-based and manipulative/"through doing" processes. Consider for example, the important role in ethics played by imagination, which, like analogy, visualization, simulation, and thought experiment, is a form of model-based reasoning. Another important theme is creativity, which is an important factor in effecting conceptual change and forging new perspectives. By exploiting the concept of "thinking through doing" and of manipulative abduction, I also illustrate some of the most interesting cognitive aspects of technological artifacts as "moral mediators." Morality is often performed in a tacit way, so to speak, "through doing," and part of this "doing" can be seen as a manipulation of the external world with the express purpose of building these moral mediators. They can be purposefully built to achieve ethical effects, but they may also exist independently of human beings' intentionality in a variety of technological entities and may carry ethical or unethical consequences.

Another fruit of my methodological perspective is a new understanding of the role of coherence in ethical reasoning: my approach provides a very general model that conveys the multi-faceted nature of comparing alternatives and governing inconsistencies. I believe the best moral deliberation results from creating or abductively selecting possible "reasons," (which, of course, may be propositional, model-based, hybrid, or a template of moral doing), then – at least ideally – choosing the most coherent set of reasons from the competing possibilities, and, finally, appropriately applying those reasons to the moral issue.

Question: How do you believe ethical and legal responsibility should be assigned when dealing with technological innovation in the world we live in?

In my book I analyze in detail the fact that we already are hybrid people, also fruit of a kind of co-evolution of both our brains and the common, scientific, social, and moral knowledge we ourselves have produced, at least starting from the birth of material culture thousand of years ago. I also maintain this knowledge production has inextricably linked us with natural and artifactual externalities endowed with cognitive functions. This condition of cyborgness complicates the cognitive status of human beings and jeopardizes their dignity by destabilizing endowments I consider fundamentally important: consciousness, intentionality and free will. I also address the problem of *preserving* these three important aspects of dignity, and I show how they are deeply connected to knowledge and, even more important, to the continual production of new knowledge and an ongoing commitment to modernizing ethical understanding.

Question: Where do you see the future of Technoethics and what approaches to research in philosophy and applied ethics do you suggest to help advance work in technoethics?

Contrary to Kant, who believed "there is not need of science or philosophy for knowing what man has to do in order to be honest and good," I contend that ethics and decision making in our technological era should *always* be accompanied by knowledge related to the situations at hand. If we want knowledge to be considered a duty, we must commit ourselves to generating, distributing, and using knowledge in service of personal, economic, and social development. Knowledge-as-a-duty is an ethical obligation not only to human beings living today but also to all those of the future.

The vital importance of knowledge means that we must use great care in its management and distribution, and as we have seen, there are several transdisciplinary issues related to this challenge: from promoting creative, model-based, and manipulative thinking in scientific and ethical reasoning to deepening the study of "epistemic" and "moral mediators"; from the interplay between unexpressed and super-expressed knowledge and their role in information technology to the shortcomings of the existing mechanisms for forming rational and ethical argumentations. A lack of appropriate knowledge creates a negative bias in concrete moral deliberations and is an obstacle to responsible behavior, which is why emphasized the potential benefits of new "knowledge communities," of implementing ethical "tools" like ESEM capacities [Earth Systems Engineering and Management] and trading zones, and of acknowledging the status of humans beings as "knowledge carriers."

These are some of the external conditions required for us to improve human dignity, but what of internal conditions? That is, how can we orient our thinking so that we are in the best position to embrace the knowledge-as-a-duty concept? Just as we must build greater knowledge about technology, so, too must we strive for deeper understanding of ourselves. I believe that increasing awareness of ourselves, comprehending how and why we think what we think is another critical component of human dignity. Before we can begin to reap the benefits of value others give to us, we must assess whether we have ascribed appropriate value and agency to *ourselves*, which brings us to the focus the link between freedom and responsibility and their relationship to some internal conditions involving bad faith and self-deception, I describe in chapter five of my book.

INTERVIEW WITH MATTHIAS KLANG, UNIVERSITY OF LUND & UNIVERSITY OF GOTEBORG

Question: How did you get involved in research in Technoethics and why?

During my law studies I became interested in the relationship between law and technology and began to specialise my studies in this field. It quickly became apparent that black letter law was not enough to identify, explain or to resolve issues caused by the increased dissemination of digital technology in our everyday lives. I was fortunate to be accepted into PhD studies and wrote my thesis on the effects of technology regulation on democracy.

Question: What are your main research interests in Technoethics and what findings in your work have you found most fascinating thus far?

In my work I have managed to combine online activism with the academic research in the fields that interest me. The main thrust of my work has been on copyright (mainly access and licensing), activists use of technology and our dependence on technology.

These areas have been particularly interesting since there has been a great deal of study in these fields and also a great deal of interest from outside the academy. A common issue of interest in all these areas is the way in which legislators attempt to regulate a technology that they seem not to fully understand.

One way of explaining this stems from the metaphors and images we use to explain technology to ourselves. The use of metaphors is important since it is a technique that enables us to understand that which is new and unknown. However the metaphors we use not only serve to help us to understand but can also act a barrier to understanding the newness of the new technology. The way in which it differs from the past.

Take for example the discussion on file sharing, or more correctly online copyright violation. Earlier motivations for limiting our permission to copy copyrighted material were based on the concept of original-copy. In other words our understanding of copyright was based upon the fundamental truth that the copy was somehow inferior to the original. However, in the digital environment this is not only untrue it may also be false since the "copy" may in fact allow the user to do more than the "copy". When a regulator fails to take this into consideration their attempts to regulate will be as flawed as their understanding of technology.

Question: Based on your co-edited Human Rights in the Digital Age, how does technoethical inquiry connect to key issues in human rights and legal informatics (i.e., copyright, free expression, censorship, access, free speech)?

When we use terms such as human rights we tend to focus on the right to life or prevention to torture and in relation to these digital rights focus more on the communicative aspects of society. At first glance this may seem less important however it is very important to keep in mind societies increased reliance on technology. The ability to access information, communicate with others, adapt information and disseminate information free from governmental, corporate or technical limitations therefore is a central role in our daily lives.

Question: One area of your work focuses on disruptive technologies. What are the ethical aspects of disruptive technologies and why is this important to society?

In my work the concept of disruptive technology is the way in which the incremental development of technology slowly invalidates previously agreed upon, and accepted, social norms. A simple example of this is the development of telephone communications. Within two decades we have moved from land line telephony to mobile telephones. This technological development has changed our communications patterns. One such example of such change is the time of day it is socially acceptable to telephone someone. Twenty years ago the caller was responsible for calling only at "acceptable" times, this has changed and today the responsibility is equally divided between the caller and receiver. One reason for this change is the receivers ability to turn his or her mobile telephone.

Understanding the way in which disruptive technologies act upon society is important both to our understanding of technology and on our ability to regulate in a society that has become dependant upon technology.

Question: In your view, how can technoethical knowledge be used to inform technology design and innovation (and diminish the influence of disruptive technologies)?

If we can create an understanding of the way in which technology effects society among users, designers and regulators of technology will enable us to identify core values in our society and develop a better understanding of the individuals role in the developing sociotechnical society. The investigation into, and development of, technoethical knowledge provide an important tool with which to study the technology upon which we depend. The goal is not to promote a luddite disdain of technology but rather a critical and aware understanding of the role of technology in society. Such an understanding will not prevent disruptive technology but it can ensure that we identify the central values which we need or want to protect.

Question: What do you consider to be the main ethical challenges in the design and development of technological innovation?

One of the main problems is that we are living in an extremely techno-optimistic society that will develop every form of technology that is possible to develop. In addition to this the techno-optimistic society believes that all problems created by new technology will be resolved by the creation of new (and better) technology. This overly optimistic approach creates a problem since it fails to recognise the responsibility of the different stakeholders (such as the user, designer, regulator, corporation).

Today the main ethical challenges lie in the conflict between the ability to access all the information available online and the desire to control such access through copyright regimes or censorship (whether state, corporate or technical).

Question: Where do you see the future of Technoethics and what research directions do you suggest to help advance new work?

The future of technoethics will have to deal with issues of design responsibility, environment and waste, digital divides, speech and censorship, privacy, openness and also issues of culture. Some of these issues are well established in technoethics while others are less explored however all issues will need to be regularly re-interpreted in relation to the developments and dissemination of new technology.

To be able to develop in this area we need a broad array of research tools and academic backgrounds. Technoethics is by its very nature a multidisciplinary field allowing for the use of a multitude of theoretical and methodological tools in order to gain a better understanding of the relationship between technology and society.

INTERVIEW WITH DR. A. PABLO IANNONE, CENTRAL CONNECTICUT STATE UNIVERSITY

Question: How did you get involved in research on ethical aspects of technology (Technoethics) and why?

I have been fascinated with technology since childhood. In fact, my interest in technology led me to pursue engineering studies for a couple of years in my country of origin, Argentina. But I was interested in many other subjects, which led me to study also mathematics, philosophy, and literature. After some years of finding my way through these interests, Argentina's political upheaval brought me to the US, where in I began attending UCLA in 1967. There, I had numerous discussions with students about ethically significant scientific and technological developments—from the 1966 cracking of the genetic code for 20 amino-acids, through changes in US agricultural practices, to the growing use of motor vehicles in transportation and their individual and social effects. These discussions prompted my growing interest in ethical aspects of technology. After receiving my BA in philosophy, I started graduate studies at the University of Wisconsin-Madison. My areas of concentration reflected my interests in ethics and technology as much as the available graduate work options permitted—ethics, philosophy of science and, as a non-philosophical area, history of science. I pursued all of these with a significantly pragmatic focus. In fact, my logic teacher once quipped that I was trying to find my way back into engineering.

After receiving my PhD in 1975, I was about to leave to teach in Texas when my dissertation adviser asked me "And now what?" I replied that I was particularly interested in problems of application. The opportunity to pursue them in a substantial manner presented itself when I was invited to teach an environmental ethics course for the Department of Philosophy and Institute for Environmental Studies at the University of Wisconsin-Madison during the Summer of 1977. In connection with that course, I started formally pursuing research on ethical aspects of technology. And I never stopped.

Question: What are your main research interests in Technoethics and what findings in your work have you found most fascinating thus far?

My main focus of research interests in Technoethics is the interface between theoretical ethics, ethical problems posed by technological developments, and policy and decision procedures involved in dealing with them. One finding which fascinates me concerns information overload, which, roughly, I understand as the excessive flow or amount of input or information which can but need not amount to knowledge and can lead to detrimental physico-technical, biological, psychological, and social effects. One aspect I find fascinating is that information overload helps understand a great variety of problems in a unified way because it occurs, in a structurally parallel manner, at the physico-technical level, e.g., when the supply of data outstrips the data storage currently available; at the cellular level, e.g., when increasing the rate of input of electrical impulses to a neuron leads to transmission breakdown; at the psychological level, e.g., when excessive stress leads to amnesia; at the social level, e.g., when the information explosion sacrifices authoritative information to easily accessed information, i.e., knowledge to speed; at the level of entire ecosystems, e.g., when the collective results of a variety of activities destabilize them. This structural parallelism may help find parallel solutions to the many problems information overload poses at those various levels.

Another finding which fascinates me is the often interactive manner in which determining right and wrong conduct, good and bad traits, justified and unjustified policies can only be done in actual circumstances. This in fact dovetails with my discussion of feedback thought in my *Philosophical Ecologies* (pp. 212-215) and with research currently being carried out on cognition, according to which people often act in order to think instead of, as standard theories of cognition have held for the past thirty years, have mental lives which take place in three discrete steps: sense, think, then act (*Science News,* October 25, 2008, pp. 25-28).

Question: Based on work in Contemporary Moral Controversies in Technology (Oxford University Press, 1987), how can moral theory be used for dealing with current technoethical issues?

That book was an initial attempt at outlining, concerning technology issues, the approach I further developed in my *Contemporary Moral Controversies in Business* (Oxford University Press, 1989), *Philosophy as Diplomacy* (Humanities Press/Humanity Books, 1994), *Philosophical Ecologies* (Humanity Books, 1999), and various articles. *Contemporary Moral Controversies in Technology* provides (and those other books further articulate), first, reasons why merely consequentialist (e.g., utilitarian) theories and merely deontological (e.g., rights-centered) theories are insufficient for dealing with policy and decision problems and issues about technology. Second, the book outlines a taxonomy of policy and decision problems and formulates the *range hypothesis*, according to which ethical problems constitute a range of problems with the following characteristics: At one extreme, individual rights carry much more weight than any other considerations in dealing with problems because, for example, natural rights are significantly and unequivocally at stake in those problems; at the other extreme, consequences carry the most weight because, for example, the very existence and well-being of reasonably good societies is at stake; and, in between, rights and consequences have less decisive weight in themselves, though, fortunately, they often reinforce each other. Sometimes, however, they appear to conflict with each other, constituting hard cases to deal with. All along the range of problems, pragmatic considerations set limits to alternatives that would otherwise have served to address the problems. This is an example of how ethical theory (in the sense of ethics as a branch of inquiry) can be used for dealing with current technoethical issues. For it leads to formulating and comparatively evaluating theories (i.e., generalized devices for formulating, clarifying, and dealing with ethical problems and issues) which, on their turn, indicate what we should primarily consider in dealing with the problems or issues. e.g., the combined theory outlined in *Contemporary Moral Controversies in Technology* indicates that we should consider the kind of situation in which given problems and issues—in the book's case, concerning technological developments—arise, whether overall societal stability is primarily at stake in the circumstances or, alternatively, whether society's stability is not at risk but individual rights are at steak, and so on.

Question: What do you consider to be the main moral controversies connected to technologies and the world we live in (e.g. environmental construction, risk-related issues at the local, regional, national, and global levels, etc.)?

One significant moral controversy of regional and, in part, global proportions concerns environmental deterioration as it relates to globalization processes. Consider the Amazonian development and its mixed socio-ecological consequences. This developments began with the 1956-1960 building of the new Brazilian capital, Brazilia, and was followed by the construction of the Transamazonic highway and multiple projects associated with it. Its consequences include a still continuing migration of millions of people from the Brazilian Northeast and Southeast, who first moved to the jungle to clear land and start small farming operations and then, with the failure of farming because the soil was leached by floods enhanced by deforestation, became part of an itinerant labor force living mostly in slum towns on the margins of the land they had cleared, and now mining for coal, gold, and iron. In the process, huge electric dams were built covering millions of acres of forest with water and providing electricity to whole sections of the jungle; an area the size of Belgium was cut or burned down (producing an enormous release of smoke into the atmosphere) for the sake of growing cocoa and raising cattle; various Native American tribes were uprooted and their numbers shrunk; many lives were lost to malaria, yellow fever, typhus, dysentery and other diseases; and, within ten years, much of the Transamazonic highway had been washed away by torrential rains, or taken back by the jungle.

Transamazonic development led to current conflicts between neighboring countries reacting to globalization processes, e.g., that about the Scandinavian paper mill operations on the shores of the Uruguay River shared by Argentina and Uruguay but, according to Argentina's claim before the International Court of Justice, in violation of accords between Argentina and Uruguay. Another is the conflict between the same countries concerning the plans by British and Australian firms to build a port terminal in Uruguay to export tons of iron and coal from Brazil. As for the connection with technology, some globalization processes (e.g., competition for global or regional markets), in combination with a variety of technologies (like those involved in the deforestation, agricultural, and mining activities previously mentioned), contribute to creating the issues, while other globalization processes (e.g., global communications through internet use) contribute to form and develop organizations which help soundly address them. The International Council for Local Environmental Initiatives (I.C.L.E.I.) and its associated truly global network, Campaña Ciudades por la Protección Climática (C.C.P.), for example, benefit immensely form global communication. So do aboriginal groups who can and do organize resistance or negotiation strategies through information sharing, email messages requesting action, cross-border and even wider coordination, and the like.

Question: How do you believe ethical and legal responsibility should be assigned when dealing with technological innovation in the world we live in?

The term "responsibility" has more than one sense; but I take it that, in this question, it is used to mean <u>obligation,</u> as in "She has the responsibility for quality control." When the obligation is moral (or ethical), it applies to moral agents, i.e., agents characteristically capable of engaging in such psychological activities as mapping out alternatives and consequences, formulating goals, identifying details of implementation, and <u>following</u> all these when acting (i.e., not merely happening to behave in accordance with them). In addition, moral agents are capable of taking the needs and interests of others seriously, as valuable in and for themselves, not merely as instrumental in attaining their own self-centered goals. In short, moral agents are natural persons. And ethical (or moral) obligations apply to moral agents to the extent that these are in control of their situation. For example, under normal circumstances, I have a moral obligation to meet a friend for dinner at a restaurant at the time I promised I would do so. By contrast, under normal circumstances I have no legal obligation to do so, because there is no law requiring people to meet their friends for dinner if they promised to do so. In other words, I incur no legal liability if I fail to meet my friend at the appointed time, though I would be morally accountable (perhaps even would be to blame) for having failed to keep my word.

Of course, not only ethical but many legal obligations—e.g., to abide by a business contract—apply to natural persons; but, by contrast with ethical obligations, legal obligations may also apply to fictional persons, e.g., corporations like Microsoft, Daimler AG. Now, though fictitious persons can be legally liable, they cannot be morally accountable because, by contrast with natural persons, they are abstract entities, hence not capable of engaging in such psychological activities as mapping out alternatives and consequences, formulating goals, identifying details of implementation, taking others' interests seriously into account, and following all these considerations when acting. To be sure, some philosophers have argued that corporations are moral persons; but this claim has been defended with question-begging arguments and flies in the face of the fact that no psychology department on Earth studies the mental life of corporations! It is the corporations' employees or members who are morally and, barring counterbalancing considerations, should also be legally responsible to the extent they influence the conduct of their organizations. For example, a corporation's upper management typically has greater moral responsibility and should have greater legal responsibility than other employees concerning the products—say, the airplanes, motor-vehicles, or pharmaceuticals—the corporation produces and makes available for public use.

At any rate, in order to determine what legal obligations should be instituted, various considerations (those previously mentioned when discussing the *range hypothesis*) should be taken into account. But this is not simply a matter of determining what moral obligations to require by law. For some legal obligations formulated by good laws are in themselves morally neutral. For example, given a great number of motor vehicles being used, laws instituting traffic regulations are good. Yet, driving on the right side or driving on the left side of the road at a time when no traffic regulations of the kind are in existence are morally neutral. (though, of course, obligations come into existence once the decision to legally require one or the other is made).

Nor is the task of determining what legal obligations should be instituted simply a matter of predicting consequences and deciding accordingly. For some significant consequences are unpredictable at decision time, e.g., those from introducing motor-vehicles in the US market in the early twentieth century. The latter was motivated by the WWII experience and the confrontation with the USSR (which did not exist until 1922!) and contributed to urban sprawl and consequently to large increases of motor vehicle use which causes about 40,000 deaths a year in the US and greater release of green house gases in the atmosphere. That is, this evidence was not available at the time motor-vehicles were introduced.

Question: What are the technoethical challenges raised by advances in business within a global society (e.g., globalization concerns, business practices, etc.) and how can these challenges be addressed?

These challenges are connected with various types of relations created or enhanced by globalization. Among these are, first, those between global business firms and other global, regional, national, or local firms (e.g., those involved in import-export trade); second, relations between global firms and countries (say, those between the previously mentioned Scandinavian paper mill built on Uruguay's Paraná River shores); third, relations between global firms, countries, and a variety of groups:

from scientists and engineers (as in the case of genetically modified potatoes produced by Argentine scientists and tested in Brazil and Chile in the recent past), through traders of iron or gold ore (as in Amazonian development), to or groups or entire countries (say, France) attempting to preserve their cultural identity, or environmentalists opposing pipeline-building and related engineering activities across Bolivia and Northwestern Argentina or elsewhere; and fourth, relations between civic sector organizations—say, The Nature Conservancy—or governmental or quasi-governmental organizations—say, the Canadian International Development Agency—and various interest groups, e.g., the International Council for Local Environmental Initiativas (I.C.L.E.I.) or its associated Campaña Ciudades por la Protección Climática (C.C.P.) (I.C.L.E.I., 2008). That is, not only are the relations involved of various kinds, but they involve a wide range of human activities and concerns—economic, scientific, technological, legal, political, and cultural.

As for addressing them, one can begin by adopting what I have called the *transcultural and transnational dialogue and interactions* approach which, in principle, makes room for a great diversity of peoples, initiatives, values, and ways of doing things. In this approach, the only common ground needed is that which furthers meaningful dialogue and interactions among the states, groups, organizations, and individuals involved. No less, but no more. The question however arises: Even if conflicting views are addressed in the previously outlined *open dialogue*, will this ensure success so that the eventual result does not end up being the mere mutual acknowledgement of differences where anything goes? The answer is no, it will not *ensure* it—failure is possible; but the approach just outlined opens up possibilities of success where other approaches have led to dead ends either because dogmatic, or too vague, or too narrow like merely legalistic or merely economic approaches.

Question: What do you consider to be the main ethical challenges in the design and development of technological innovation in the world we live in?

One main challenge is for researchers, ethicists, and policy makers is to proceed with an awareness of the fact that there are likely to be long term (or even intermediate term) unexpected consequences of innovation, and avoid technology policy making tunnel-vision. This can be done by providing reasonable, practical leeway to adapt to new situations before it is too late because of an excessive commitment to one technology or way of doing things. I already mentioned a significant example—introducing motor-vehicles in the US market in the early twentieth century and its combination half a century later with the Interstate Highway System. Another example is that of the US agricultural practices introduced during the Nixon administration to curb the relentlessly increasing cost of food. Farmers were given subsidies to grow grains (but not other crops) from fence to fence on their land, which undermined the raising of cattle in those lands, which made it necessary to use artificial fertilizers where these used to be provided by rotating cattle and crops. The primary focus on this approach left little room for alternatives and resulted in an agribusiness system whereby, arguably, misnutrition increased in trying to prevent malnutrition. By contrast, a large grain and meat producer country like Argentina by and large, and successfully, follows an eight year rotating system which, roughly, devotes large tracks of land to raise cattle for five years (thus fertilizing the land) and then to grow all kinds of crops for three years.

Another challenge concerns the designing and development of technological innovation in ways that do not undermine people's sense of personal identity and human community. This challenge is posed by a range of technologies—e.g., genetic, computational—which aim at providing technological enhancements. It has been often stated that these enhancements are morally permissible only insofar as they are therapeutic—e.g., genetic modifications to treat sickle-cell anemia and the "bubble boy" immune deficiency syndrome—; but not when they go beyond that—e.g., as genetic enhancements for athletes or, as it is commonly called, *"gene doping,"* or *as genetic memory enhancements* aimed at improving other competitive performances, say, in SAT's taking, and the like. Beyond the scientific problem of developing reliable tests for gene doping and its cognates, there is the not merely scientific problem of distinguishing between therapeutic and merely aesthetic or other uses of genetic modifications—e.g., is a modification aimed at helping attain normal height to a child who otherwise would be extremely short therapeutic or aesthetic? *Also, more crucially, there is the problem of delineating what modifications would be so significant as to alter our personal identity, or humanity (making us transhuman?), and how to evaluate them and the policies, decisions, and social-conventions changes that would lead to or be involved in them.*

Question: Where do you see the future of Technoethics and what research directions do you suggest to help advance work in this domain?

I believe technoethics has a bright future so long as it inserts itself in real-world, increasingly global tensions and seeks new forms of balance between ideas that conflict and people who contend but must live with each other. In this regard, the previously mentioned transcultural and transnational dialogue and interaction approach should help. While the conceptions of ethics and in particular, of technoethics associated with the conceptions of philosophy as *the study of large unsolvable problems*, or as *a dialogue about fundamental problems*, or as *the task of underlaborers* of the sciences, are too narrow and constrictive. So are mere interdisciplinary—hence merely academic—approaches, which by their very nature fail to engage the larger society. For one, they do not help us deal comprehensibly with issues concerning the uses of technology, which to be sure are not merely academic. Nor do they help us deal with the networks of ideas or outlooks that range not merely over the community of philosophers, or even over the wider community of intellectuals, but over society at large. The approach I have suggested is free from these limitations and may help us, humans, raise ourselves above our conflictive times.

INTERVIEW WITH NEIL C. ROWE, U.S. NAVAL POSTGRADUATE SCHOOL

Question: How did you get involved in research on ethical aspects of technology (Technoethics) and why?

I am interested in technoethics because I teach at a military university, military organizations are heavy users of technology, and ethics issues arise in many areas of military operations. Most of my background is in the artificial-intelligence area of computer science. I have been particularly interested lately in the building of models of deception, and deception is a topic with many related ethical issues. Despite many years of work, I feel that important progress can be made on understanding it with new taxonomies and quantitative models. Recent interest in deception has been prompted by the relative ease of deception in cyberspace (witness cyber-scams and cyber-attacks), wherein deception can be automated, permitting mass production of ethical dilemmas.

Question: What are your main research interests in Technoethics and what findings in your work have you found most fascinating thus far?

Military organizations address many technoethics challenges today because they have a wide range of technology for imposing force on people and objects, and some of the technology can be lethal. Conventional weapons like guns and bombs raise technoethics issues, but they improve only slowly and a body of well-established ethics guidelines governs their use. There are fewer ethics guidelines for new kinds of weapons. However, often there is consensus that the weapon should not be used at all. Nuclear, chemical, and biological weapons are prominent examples of weapons whose use is generally considered unethical because they are difficult to control and avoid hurting civilians disproportionately. I have argued that cyber-weapons, tools that attack computer systems of an adversary, are unethical for the same reasons. International treaties and conventions, with adequate enforcement machinery, are our best recourse against hard-to-control weapons, and should be used against cyberweapons with high priority.

Contrary to popular belief, weapons are only a small part of military activities: Most of it is posturing in various ways. So there are many issues in military technoethics besides those of weapons. Improvements in computer networking have permitted military activities to be managed at an increasingly great distance from their targets. We saw this in the U.S. occupation of Iraq which began in 2003 with much initial micromanagement from Washington. For instance, Washington experts would pour over satellite images and try to find buried explosive devices without much success. In general, increased top-down control has not led to better results, and the trend is now to give more control and decision-making to local commanders. It has the side effect of permitting better ethical judgments, since the closer you are to a situation, the better your access to the facts you need.

Question: Cyber-terrorism is an important topic in Technoethics. What are the main ethical concerns in cyber-terrorism and counter strategies (i.e., cyber attacks and counter attack strategies)?

Cyberterrorism could raise technoethics issues in the future. So far we have not seen any real examples of it. Important cyber-resources are well defended, and cyberterrorists would have even more difficulties in controlling and targeting their cyberweapons than military organizations would. So it will be hard to get near the same impact with a cyberweapon than by, say, blowing up a computer. Thus proactive responses to threats of cyberterrorism, such as violating the privacy of citizens to search for clues, seem unjustified today. Passive methods such as patching bugs in software should be sufficient protection for now.

Question: What are the main ethical concerns in deception plans for defense in Cyberspace? How should these concerns be addressed?

Defensive deception in cyberspace might also seem to raise ethics issues, but based on my research I am doubtful these issues are serious. Defensive deception methods are generally passive techniques such as fake files, false error messages, and delaying tactics that only trouble people with malicious intent for your computer systems. Incomprehensible files, obviously-wrong error messages, and unexplained delays are not uncommon anyway on normal computer systems. However, some more proactive defensive techniques like logging into adversary computer systems to plant monitoring devices could raise ethical issues. Generally speaking, computer technology that provides power but minimally raises ethical issues is desirable – and defensive deception is an example.

Question: What are the main technoethical challenges in the military? Who should be responsible for controlling advanced military technology and what ethical aspects need to be addressed?

In general, most ethical challenges with technology concern its distancing between actor and effects. To use a common military term, technology is a "force multiplier":

It permits you to exert more power at a greater distance. So when you can push a button to turn off a power grid or send a missile to a destination, these serious actions have been made overly easy and thus susceptible to abuse. This does not mean that technology is necessarily unethical, however. Added power can be used for added good as well as added evil. If used carefully by enlightened people, the power of technology can do many great things.

Many ethical problems arise because people are unaware what the technology actually does, either because they misunderstand its direct effects or they do not see its indirect effects. This, for instance, is a big problem with cyberweapons: Effective cyberweapons must usually be kept highly secret until used, because most countermeasures are simple software modifications that can be found easily given enough time. But if hardly anyone knows what the cyberweapons do because they are secret (and these weapons require extensive software expertise to understand anyway), they can more easily hurt people. Hurting people with military technology includes less dramatic acts than explosions, like violation of privacy. It also includes plagiarism and stealing of ideas which have become easier and more common with networking technology.

Poor understanding of the effects of technology is exacerbated by over-optimism about technology. Military organizations, like most large bureaucracies, like to periodically have something new about which they can get excited, whether it fulfills a genuine need or not. Military organizations also feel pressure from competition with other military organizations. This creates pressure to use new and untested technology before it is debugged, and can result in high rates of problems, both technical and ethical. Similarly because of diverse influences, bureaucracies like to create complex objects and procedures in an attempt to appease all influences. So military software systems tend to have overly large numbers of requirements and features, and tend to be too large for their function. Bugs are intrinsic to large software systems, and their number is roughly proportional to the square of the size of the system. So many military software systems are unreliable, and using them in matters of life and death may be unethical.

Even when military organizations and their personnel understand their technology, there are still many ethical questions because military decisions may need to be made quickly without all the facts. For instance, suppose a military is suddenly under cyber-attack from what appears to be a single country: Is it ethical for that military to use cyber-counterattacks? This could be the only way to stop dangerous attacks immediately, but could have severe repercussions on the innocent civilians in that country, and the country itself might be the innocent victim of hackers. Even turning off the country's Internet connections could have severe repercussions. So complex ethical tradeoffs must be assessed between the cost of doing nothing and the cost of doing too much.

We should actually welcome the new ethical problems raised by technology because they encourage us to reexamine our principles. The world has changed so much in the last hundred years that many ethical judgements from then have little relevance today. This suggests that our laws and religions need reexamination. For instance, in a world in which crimes can be committed by people who are physically remote and without values in common, the notion of local policing, the basis of our crime management of the last five hundred years, is now invalid. Similarly, ubiquitous advertising in modern life makes nonsense of ethical pronouncements against lying. Still, some concepts remain valid, such as the notion of unjust enrichment which is still a good guide to determining what is unethical in Internet business activity.

Question: Where do you see the future of Technoethics and what research directions do you suggest to help advance new work?

Technology could actually help ethics in some ways. We can use the ability of computer technology to manage large amounts of data to help us manage ethical decisions. For instance, consumers are exposed to many advertisements with little ability to judge their claims. Data-mining computer software could collect information about the businesses, such as objective ratings of the quality of their products, the history of legal actions against them, their labor-relations records, what governments they have interacted with, and the ethics of their business in general. With computer aid, we could construct a set of rules reflecting our weighted ethical judgements about these businesses. We could run those rules automatically to advise us on making informed ethical decisions.

INTERVIEW WITH DR. A. ANDERSON, UNIVERSITY OF PLYMOUTH

Question: How did you get involved in research in Technoethics and why?

My background is in the discipline of sociology and I have a longstanding interest in the role of the media in communicating environmental risks (see, for example, *Media, Culture and the Environment* (Taylor & Francis, 1997). Given the increasing attention being paid to nanotechnology in early 2000, I developed a particular interest in examining the role of the news media in framing its social and ethical dimensions. In 2004 I was awarded a grant from the Economic and Social Research Council (with Alan Petersen and Stuart Allan) to examine national press portrayals of the potential benefits and risks of nanotechnologies. This was followed in 2006 by a grant funded by the British Academy (with Alan Petersen as principal applicant) to investigate how scientists and policymakers portray the benefits and risks of nanotechnologies, as applied to medical and environmental sustainability. These were among the first studies to be undertaken on the social aspects of nanotechnologies in the UK.

Having previously examined media framings of biotechnology, I became especially interested in exploring the question of whether the debate over nanotechnology is likely to become a re-run of the crisis over genetically modified (GM) food and crops. Whilst some commentators claim that nanotechnology raises no new ethical issues, evidence suggests that it possesses some distinctive features, which clearly set it apart from other case studies of technological development (see Wood et al. 2008). In particular, the convergence of nanotechnologies with other technologies poses important new social and ethical challenges (see RS/RAE 2004). Such challenges arise from the broad range of industry sectors, materials and devices covered by the umbrella term 'nanotechnology'; which is neither a new nor a single technology (Wood et al. 2007). Moreover, many areas are still at a relatively early stage of development and there is considerable uncertainty about the potential effects of nanoparticles on humans and the environment. The trans-disciplinary nature of nanotechnology and its diversity of applications make it a distinctive new area for the study of technoethics. Accordingly, 'nanoethics' is rapidly becoming an important new area of academic study, evidenced by the new journal of the same name and an increasing range of books published on this topic within the last few years (e.g. Alhoff et al. 2007; Alhoff & Lin 2008; Hunt & Mehta, 2006; Schummer & Baird, 2006).

Over the past 10 years nanotechnology has become the most rapidly rising sector in the knowledge based economic infrastructures of OECD countries (Throne-Holst & Stø 2008). The 'nano' prefix in 'nanotechnology' derives from the Greek word 'nanos' which translates as dwarf. One nanometre is one billionth of one metre, which is tens of thousands of times smaller than the diameter of a human hair. As with biotechnology, nanotechnology involves control over life (in the form of nanomedicine). However it also encompasses nano-manufacturing with applications in a range of consumer products such as sunscreens, cosmetics, scratch resistant paints, textiles and sports equipment, with additional applications in electronics, engineering, and the military. There are currently in excess of 600 products consumer nano-products available on the market, yet public awareness of these developments is low and there are serious concerns about a lack of adequate regulatory control (Macoubrie 2006; Michelson & Rejeski 2006). The potential dangers associated with technology can be under-estimated, as in the case of asbestos or thalidomide, over over-played, so it is important that both risks and benefits are rigorously investigated.

While much of nanotechnology's potential has yet to be developed, it has already attracted considerable controversy. Some bioethicists, as well as campaigning groups such as the Canadian ETC Group and Friends of the Earth have called for a moratorium on all research until the apparent gap between the science and ethics is closed (Mnyusiwalla et al., 2003; ETC, 2003). Ethical issues arising from nanotechnologies have tended to be perceived from a utilitarian approach to 'benefits' and 'risks' (Wilsdon & Willis, 2004). This shifts attention away from unpredicted or unknown effects (which are, by their nature, hard to anticipate) and neglects to provide publics with a real chance to express their perceived social and ethical concerns. This approach assumes that identified problems can simply be dealt with through further research and enhanced regulatory procedures, thereby neglecting the social complexity of nanotechnologies, the exercise of power in problem definition and the involvement of vested interests in certain 'framings' of issues (Anderson et al. 2008).

Question: What are your main research interests in Technoethics and what findings in your work have you found most fascinating thus far? Please give a couple of examples from your research on technology controversies (e.g., nanotechnology, biotechnology, etc.).

We have been particularly interested to explore why there has been an absence of reference to particular actors and issues in news media portrayals of nanotech (such as economic, ethical, theological, and legal issues) and whether particular framings are contingent on certain events or periods in the policy making cycle. The highest proportion of all our articles, in a content analysis of UK national newspaper reporting of nanotechnology between 1 April 2003 and 30 June 2004, featured science-related frames and the most prominent stakeholders in the coverage were scientists (see Anderson et al. 2005; Anderson et al. 2008). Most of the sampled newspapers devoted more coverage to benefits outweighing risks, though there were some interesting differences among newspapers. The scientists and policymakers interviewed in the BA funded study, while acknowledging the many uncertainties about the risks and their assessment, were mostly optimistic about the benefits of nanotechnologies (see Petersen and Anderson, 2007; Petersen et al. in press). Our research suggests that there are a number of parallels that can be drawn between nanotech and previous controversies over other 'emerging technologies' such as biotechnology and stem cell research. The heightened expectations surrounding nanotechnologies are similar to those accompanying the emergence of biotechnology.

However, from the start a concern with societal implications has been at the forefront of nanotechnology development, given the anxieties that a GM style public backlash could take place. Our research suggests that negative frames have tended to have featured more heavily in early press coverage of nanotechnologies, particularly in the UK, compared with the first two decades of biotechnology coverage (see Anderson et al. 2005; Anderson et al. 2008) However, in overall terms the emphasis upon economic progress has been especially prominent, particularly in the US. From the start the degree of importance attached to societal benefit and consumer applications, itself part of a broader movement towards encouraging scientists to deliver more commercially exploitable products, can be seen, to some degree, to distinguish it from other scientific developments (Gorss & Lewenstein, 2005). As nanotechnologies gain a higher profile within press coverage, I expect the ensuing debate may become much more sharply polarised, possibly reflecting a shift from predominantly scientific frames to an increasing concern with ethical implications (see Anderson et al. 2008; Anderson et al., in press).

Question: What are the potential health and safety risks with nanotechnology applications (and other controversial technologies) and who is responsible?

In addition to a range of potential benefits there are a number of possible risks associated with nanotechnologies. The environmental, health and safety impacts are among the most prominent concerns. To date there is relatively little available data on the risks, though it was recently announced that the EU are to fund a 2.6 million study on nanoparticle safety commencing in 2009.

There are concerns over the potential health hazards of nanoparticles (these can enter the body through the lungs, intestinal tract and skin, although the skin forms a more complex barrier (Hoet et al., 2004). We are already exposed to a variety of natural nanoparticles; in fact, these have existed naturally since the beginning of the Earth's history. Since natural nanoparticles have been found to be toxic under certain conditions, this raises the fear that manufactured nanoparticles could also pose a danger (Handy et al. 2008). There is considerable uncertainty about the effects of nanoparticles on the environment, but some emerging evidence suggests that they may cause biological harm (Colvin 2003; Preston, 2006). Concerns have been raised over the environmental impacts of nano-manufacturing processes, especially those which involve high amounts of water. Toxicological evidence from animal and cell studies suggests there may be the potential for harmful effects on our lungs and immune systems if exposure occurs at high levels (Handy and Shaw 2007). There are claims both that nanotechnology can be utilized to kill cancer cells and that carbon nanotubes pose a cancer risk. As with other technology developments, such as stem cell research, there is also the danger of potential misuse in medical research.

In the light of previous controversies over GM food and crops, there are particular fears over the possibility of humans becoming exposed to nanoparticles via the food chain. Friends of the Earth Australia have raised particular concerns over food safety issues. According to the pressure group there are over 100 food, food packaging, and agricultural products containing nano ingredients presently on sale internationally in the absence of mandatory food labelling (FOEA, 2008).

Question: What are the rights of individuals in affecting nanotechnology applications and what responsibilities do technology developers and leaders have to protect society from the risks of nanotechnology advancement?

In the case of nanotechnology I think there is a real opportunity to address social and ethical issues early on in the research and development process. Mindful of the potential for a crisis in public confidence to occur, increasing emphasis has been placed upon engaging publics during the early stages of technology development, especially in the UK. However it is questionable as to how far efforts to move public engagement 'upstream' have genuinely opened up wide-ranging debate about the implications of nanotechnology among a diverse range of publics (see Anderson et al. in press; Petersen et al. in press). Given the complexity of the science, together with a general lack of awareness of nanotechnology, it is difficult to see how knowledge can be built completely from the bottom-up (Wood et al. 2008). Nevertheless, risk analysis must be conducted in as inclusive and transparent way as possible. A preliminary study in the US suggests that industry and university scientists are the most trusted sources of information for the public (Scheufele et al. 2007). As key authoritative voices in the nanotechnology debate industry and scientists play a crucial role in the framing of issues, particularly in relation to the benefits and risks of technologies. Along with policymakers they have a responsibility to ensure that publics are reliably informed about the nature and potential impact of innovations in this field.

Nanotechnology thus raises important issues of responsibility, accountability and ownership. In my view key questions that need to be asked include: What do people need to know about the potential risks? Who will be most affected by potential risks? Who is likely to benefit most from particular applications? And crucially, as Bruce Lewenstein points out, who should decide what becomes defined as a social and ethical issue in the first place? (Lewenstein 2006).

Question: What do you consider to be the main ethical challenges in the design and development of technological innovation?

In my opinion the complexities and fragmentation of the nanotechnology field pose particularly tough challenges for regulation and there are concerns that some nanomaterials could completely bypass testing and safety evaluation. The Nanotechnologies Industries Association has worked with the Royal Society, Insight Investment and the Nanotechnology Knowledge Transfer Network to develop a voluntary Code of Conduct for all involved in developing and exploiting nanotechnologies (Responsible Nano Code 2008). Particularly difficult issues are raised by scenarios where liability among existing stakeholders cannot be determined and the original manufacturer cannot be traced. Corporations will need to balance self-interest against the need to act in an ethical responsible manner. There will need to be greater investment in strategic risk research in order to consider indirect, cumulative and synergistic effects (Lee and Jose 2008).

Question: Where do you see the future of Technoethics and what research directions do you suggest to help advance new work?

There is increasing interest in the ethical dimensions of technological developments and growing acknowledgement of its importance. I think Technoethics will continue to gain greater prominence as a significant interdisciplinary field of study in the years to come. In terms of new research directions, I think that the processes through which risks are communicated deserves much greater attention. Discussions about the implications of new technologies rarely devote much attention to the dynamics of news reporting, yet precisely how the media frame the nature and range of issues associated with emerging technologies is of critical importance since it has the potential to legitimize certain definitions over others (Anderson et al. 2005; Anderson et al. in press). Science and technology exist within a social context and the power to define what is regarded as a social and ethical issue is crucial. In my view, much greater attention needs to be paid to the role of the media; in particular, how news is socially constructed and how different stakeholder groups seek to establish their preferred representations of issues in ideological terms. In the competition to secure favourable media coverage levels of source access to the media reflect wider power differentials within society. Moreover, power may be exercised to suppress as well as publicise controversial news items. By framing scientific issues in particular ways the news media help set the preferred terms in which questions about potential problems and solutions are defined. These include questions such as who is likely to own or have access to these technologies and who is likely to be advantaged or disadvantaged in relation to them? There is an urgent need to reflect upon how journalists may better communicate the uncertainties and risks of emerging technologies in an ethically responsible way.

INTERVIEW WITH DR. LYNNE ROBERTS, CURTIN UNIVERSITY OF TECHNOLOGY

Question: How did you get involved in research in Technoethics and why?

In the mid 1990's I commenced work on my PhD examining social interaction in virtual communities. I was fascinated by people's experiences interacting online, at that time predominantly using text-based computer-mediated communication. In amongst findings of positive social interaction, networking, friendship and love were tales of flame wars, harassment, deception, gender-switching, obsession, stalking and surveillance. The potential for anonymity or pseudonymity, and the additional surveillance possible online were resulting in some users engaging in a range of negative behaviours that had both online and offline consequences for their victims. This sparked my interest in possible unintended consequences of information and communication technologies.

During this period I also collaborated with Professor Mac Parks from the University of Washington on a research project on gender-switching online (Roberts & Parks, 1999). Gender-switching highlights the unprecedented control over the construction and presentation of identity afforded by information and communication technologies. We found a wide range of beliefs about the acceptability of presenting an identity online that varies from the offline self, from a belief that it is dishonest and manipulative through to acceptance and embracing the affordances offered by the technology for identity exploration.

My research at this time was conducted online using a mix of online interviews and online surveys. I was engaged by the early debates on the ethical conduct of research online. Online researchers were struggling with issues such as concepts of public versus private space online, fair use and copyright issues relating to online postings, confidentiality (including whether pseudonyms, as well as person identities, needed protection), and how to obtain and document informed consent online. Based on my developing interest and experience in this area, I began writing on ethical issues in conducting research online (Roberts, Smith, & Pollock, 2004)

Several years later while working as a research fellow with the Crime Research Centre at the University of Western Australia I became interested in cybercrime, once again focusing on the unforeseen uses of information and communication technologies.

Question: What are your main research interests in Technoethics?

Using Luppicini's (2009) typology of key areas in technoethics, my research interests fall squarely into the area designated as 'internet ethics and cyberethics'. Building on my earlier experiences outlined above, my two core technoethics research interests remain online research ethics and cybercrime. I have continued to publish in the area of online research ethics (Roberts, Smith & Pollock, 2005/2008; 2008) and am a reviewer for the Human Research Ethics Committee at my university. In examining cybercrime I have published papers on cyber identity theft (Roberts, 2009b) and cyberstalking (Roberts, 2008) and currently am supervising students completing research projects on cyberbullying and illegal downloading online. However, increasingly my focus is on 'cybervictimization', the impact of interpersonal cybercrimes on individuals (Roberts, 2008a, Roberts 2009a), an area I hope to pursue in the future.

Question: What is cyber-crime and what types of computer-mediated interpersonal crimes exist today?

Cybercrime refers to the use of information and communication technologies (ICTs) for criminal purposes. ICTs can be used as the mechanism for organizing and committing criminal activity and also as a means of protecting criminals against detection. Networked computers have enabled the emergence of new types of criminal activity and new ways of conducting existing criminal activities, expanding both the range of criminal opportunities and potential victims.

Computer-mediated interpersonal crimes are criminal activities conducted online that take the form of an 'assault' against the individual, their integrity or reputation. This includes cyberharassment, cyberbullying, cyberstalking and the online sexual exploitation of children Cyberharassment and cyberbullying typically involve online messages designed to harass, humiliate and intimidate and may include threats, insults and teasing. Increasing multimedia capacity extends the range of material from text based messages to videos, images and voice. In addition, cyberstalking may include threats, harm to reputation ('cyber-smearing'), damage to data or equipment and attempts to access confidential information and computer monitoring. Key areas of concern in relation to the online sexual exploitation of children are the manufacture, transmission and viewing of pornographic child sex imagery and the sexual grooming and solicitation of children online. There is debate over whether these interpersonal cybercrimes are best conceptualised as new forms of deviant and/or criminal behaviours or are simply an extension of offline behaviours. What is clear is that the Internet provides a wide range of opportunities for individuals to interact with strangers, expanding the potential 'pool' of victims, and also the potential audience, for these types of harassing activities.

Question: What are the ethical and legal challenges in dealing with cyber-stalking and cyber-victimization?

Cybercrimes are relatively new crimes and responses to cybervictims are still developing. It is important that a range of services (including crisis intervention, counselling and advocacy) are available to support those victims who require assistance. In order to determine service provision requirements and appropriate responses to support current and future cybervictims significant further research is required into the nature and extent of interpersonal cybercrimes and the impacts of cybervictimization.

Interpersonal cybercrimes also raise new issues for the legal profession in terms of the 'harms' experienced in the absence of the physical presence of the perpetrator. No longer is physical proximity required for harm to occur. Effective law enforcement and legal responses to interpersonal cybercrimes are dependent upon the formulation or amendment of laws to recognise the harms that can result in the absence of physical proximity. Further, the potential cross-jurisdictional nature of these crimes requires coordination and co-operation between jurisdictions. A continued focus on international cooperation is required.

A further ethical challenge in dealing with interpersonal cybercrimes is balancing the competing rights of individuals against civil liberties and privacy of all Internet users. Increasingly, technological 'solutions' to cybercrimes are proposed. Ideally, attempts should be made to identify and evaluate the potential effects of each new technological 'solution' prior to implementation, so that a considered judgement can be made of the overall costs to all Internet users and the benefits that might accrue in terms of protecting some individuals from cybercriminal activities. Without sufficient independent scrutiny there is a danger of 'function creep', where a solution may expand to cover areas not originally intended, potentially infringing upon the rights of other Internet users.

Question: What do you consider to be the main ethical challenges in the design and development of technological innovation?

The need for technology assessment as part of the design and development process has long been recognised with work in this area commencing in the 1960's. However, examining the ethical and social consequences of technologies as part of the design process has received limited attention (Palm & Hansson, 2006). Even where an ethical technology assessment is conducted as part of the design and development process, it is unlikely that accurate predictions of all future uses of a technology will be made, highlighting the need for ongoing ethical consideration (Palm & Hansson). Indeed, it has been argued that ethical problems increase as technological revolutions progress ('Moor's Law'; Moor, 2005). Wider dispersion of a particular type of technology means that not only are more people affected by the technology but that new opportunities emerge for the use of the technology, including for criminal purposes. Moor described the computer/information revolution as a technological revolution that has passed through the 'introduction' and 'permeation' stages and that by 2000 has entered the 'power' stage. New ethical issues continue to arise and are often not covered by current policies, indicating the requirement for ethical consideration beyond the development stage. A key question to be addressed is given the limited resources available, at what level should ethical technology assessments be focused? Should the emphasis be on the design process of individual applications/ technologies (many of which may never 'get off the ground') or on broader 'classes' of technology that emerge as problematic over time? In relation to cybercrime, it can be expected that the form of cybercrimes will continue to change over time as new ICTs and applications emerge.

INTERVIEW WITH DR. DEB GEARHART, TROY UNIVERSITY

Question: How did you get involved in research in Technoethics and why?

I have been working in the field of distance education/distance learning for over 23 years. As our field moved into the electronic age I found that students did not know about the ethics of working with technology. I started research on the subject. As a distance instructor and a distance program administrator, I was concerned about students' ethical behavior, particularly in the areas of cheating and plagiarism.

Question: What are your main research interests in Educational Technoethics and what findings in your work have you found most fascinating thus far?

I started looking at student's perceptions on use of technology and what is ethical. One phenomenon I looked at was "psychological distance" where students using technology do not see the harm done to others (or to themselves) by their actions using technology. Students see no problem in cutting and pasting information on the Web into assignments. There is no concept of copyright violation, plagiarism, etc. I recommend in one study that technology ethics should be taught in school as soon as technology is used and the level of ethical behavior be further developed as students work through their education. My continued research has been on expanding the concept of ethical behaviour with technology and includes software piracy, use of moral language and behaviour, privacy and confidentiality, understand the concept of doing no harm in research, teaching and everyday behaviors.

Question: Tell me about the higher education act and the importance of educational technoethics related to teaching and learning (e.g., student authentication issues)?

This is my current research area as I deal with it on a daily basis. Even before the language was introduced in the Higher Education Act, Troy University was working in conjunction with Sexurexam to develop the remote proctor which the institution is currently being implemented in the eCampus program. I have been doing presentation on student authentication and the ethical and technoethical issues. This will be the topic of my next research article.

Question: What would you say are the main technoethical challenges in online and traditional classrooms contexts?

The increased use of online exams and quizzes are a technoethical issue both in the classroom and online. Although I feel that students who cheat is still a small number it is time consuming for faculty and administrators to deal with. Students continue to work at finding ways to get around security measures for online exams. Another area of concern is how students use technology to obtain information in non-ethical ways which causes both plagiarism and copyright issues. A question that educators and researchers need to address is at what point does the way our students learn become affected by technology to where we no longer teach in the manner we always have. Should students be expected to learn dates in history, know math, or should we just looking it up online? Where does this technology take us in the future of teaching and learning?

Question: How do advances in educational technology affect access to new educational resources and the growing digital divide?

As someone who has been in distance education/distance learning since before the "technology revolution" I have been concerned about the digital divide for a long time. We will always have groups who will not have access to technology for educational purposes; groups like incarcerated students, migratory workers and their families, rural students with no access to Internet, lower income families. We are broadening the divide when we are closed minded to the needs of others, globally, who do not have technology access.

Question: What do you consider to be the main ethical challenges in the design and development of technological innovation in education?

I think that determining ethical standards with technology use and then providing education of these standards is a main challenge for this research area. I did a little looking for ethical standards of technology use in education while writing my last article and found little information dealing specifically with ethical use of technology. Technology advances are so quick and we are not keeping up with standards of use in education. What happens then is that our students develop poor technology use habits and this lack of technology ethics greatly impacts them in higher education and later in the workplace. We are seeing an increase of plagiarism in courses from our students.

Question: Where do you see the future of educational technoethics and what research directions do you suggest to help advance work in this area?

Being in higher education and in distance education in particular, I think that the research in technoethics needs to deal with technology and learning, is it effective? And with the new hot topic of student authentication.

INTERVIEW WITH DR. MAY THORSETH, NTNU NORWEGIAN UNIVERSITY OF SCIENCE AND TECHNOLOGY

Question: How did you get involved in research in Technoethics and why?

The initial reason why I got involved in ethical aspects of technology can be traced back to 2002. As a director of Programme for Applied Ethics (see http://www.anvendtetikk.ntnu.no/main.php) at my university NTNU I arranged a PhD course on Internet and Ethics. Through this arrangement I learned to know prof. Charles Ess

(Drury University) who was doing parts of the teaching of the course. By that time my own research was not involved with any technological aspects at all. However, later - in 2004 - I was invited the North American Computing and Philosophy Conference (NA-CUP) in Pittsburgh, Carnegie Mellon University. Here I met some very prominent scholars in this broad field of computing and philosophy, among them Robert Cavalier. He had done research on online polling that I found very interesting. I myself was already involved in a huge multidisciplinary research project called Democracy Unbound (see http://people.su.se/%7Efolke/index.html) at that time. My main research interest was on worldwide democratic deliberation. My interest for looking at the internet as a venue for worldwide democratic deliberation started here.

Question: What are your main research interests in Technoethics and what findings in your work have you found most fascinating thus far?

My main interest in this field has been to research the internet as a possible venue for democratic deliberation on a global scale. An interesting case in question has been online polling (cf. James Fishkin and others). What I have found most fascinating is the Janus face of it: on the one hand there are interesting reports that it is possible to carry out this kind of polling for the purpose of qualifying public opinion ahead of elections, for instance. The PICOLA project is interesting in this respect (see http://caae.phil.cmu.edu/picola/index.html). On the other hand there are also reports on polarisation and filtering (cf. Cass Sunstein). Another fascinating aspect of this is the potential for empowerment of minorities on the one hand, and - partly along with it - the potential for mobilising in the online works for (political) actions in the world outside the net.

Another fascinating aspect of the technology is the fragile and even blurred distinction between the real and the virtual. My own interest in worldwide deliberation and multicultural conflicts has made me realise that the virtual realities that the internet offers has not as yet been fully utilised in this field of interest.

Question: What are the dangers of technology innovation within a context of globalization and multicultural conflict? How can these dangers be addressed within a technoethical framework?

The dangers of technology innovation are many. The most pervasive is perhaps the fact that the more expertise that is developed, the more will democracy be jeopardized because most people cannot but trust the technology while being gradually more dependent on it. A technocracy might be lurking in the background. Unless we develop cultures with a high degree of transparency we might fall victims to our own best intentions: to facilitate free and un-coerced communication among citizens. Daily Me (Sunstein) and hate groups are other risks in question: the risk that people do not engage in but a limited set of dialogues in society and only get the information they want to; on the other hand the risk that extremists might act quite freely and even have their already extreme viewpoints strengthened. This is directly related to the threat of terrorism offline.

Question: In your opinion, how can global online communication and deliberation be used within the context of technoethical inquiry to address key areas? (e.g., assigning corporate responsibility, participatory decision making, virtual organizational work, conflict resolution, etc.)?

In my latest research I have started investigating the possibilities of utilising the potential for counterfactual thinking that the virtuality made possible on the internet might offer. More concretely I have in mind online societies like Second Life and the like. The philosophical problem of the public – the need for improvement of the methods and conditions of debate, discussion and persuasion raised by John Dewey (The Public and its Problems, 1927) - calls for improvements of methods for deliberation. However, considering the limitations of online polling, blogs and the like, I think we have to look for new methods. In my recent publication "Reflective judgment and enlarged thinking online" (http://www.springerlink.com/content/e1vr2554g732v805/) I have discussed how there might be an interesting philosophical link between online virtuality and the Kantian concept of reflective judgment. Put simply, the problem of the public is here formulated as a problem of how to make people overcome the limitations that contingently affect our own judgments.

Question: What do you consider to be the main ethical challenges in the design and development of technological innovation within a global context?

In my view the main ethical challenge is to make technological innovation facilitate people's possibilities to overcome their own limited subjective conditions in how they consider themselves (and others). Within the debate on deliberative democracy some of the difference democrats have argued why different forms of communication should be allowed in deliberative processes (e.g. I.M. Young, on storytelling, testimony and rhetoric). This has often been part of the discussion on empowerment, and how to make underprivileged minorities capable of participation. This is of course an important concern, among others, relating to the digital divide. An even more serious challenge is the fundamentalism that is made possible through this technology, and how to find a balance between preserving freedom of speech against the need for control in order to protect the same ideal.

Question: Where do you see the future of Technoethics and what research directions do you suggest to help advance new work?

I believe technoethics should have a primary concern for the way people use the technology. Studies on e.g. virtuality do, of course, have to be released from applied contexts to some extent. At the same time I think it is urgent to see to it that the different research groups and research communities communicate with each other, and mutually inform each other on their findings. We need to inform those who are not themselves deeply involved in the technical aspects of this technology how it actually works. This is urgent because the policies ought to be democratically based. In order for non-technological scholars to be able to have a qualified opinion of the ethical and political choices that are developed and being built into the technology we need to have a dialogue between more technologically and non-technologically oriented scholars. The same holds for the dialogue between the research society and the remaining society, as well.

Glossary

Anti-Spyware Software: Anti-spyware software is software that regulates which software is permitted on the users computers and which is not.

Applied Ethics: Applied ethics is a branch of philosophy dealing with the ethical values within society. It focuses on practical real-life ethical dilemmas such as abortion, racism, and sexism.

Artificial Intelligence: Artificial Intelligence (AI) describes computer programs and devices which simulate human functions and activity. It also refers to the research program aimed at designing and building intelligent artifacts.

Artificial Life: Artificial Life refers to a broad research program focusing on the structure and processes of life (i.e., self-organization, self-reproduction, adaptation, evolution).

Artificial Moral Agent: Artificial moral agents refer to moral agents whose ontology is detached from the natural world. The possibility of artificial moral agency is a controversial topic of debate.

Assistive Technology (AT): Assistive technologies (also called *adaptive technologies*) are devices used to improve the functional capabilities of individuals with physical disabilities. (e.g., computer screen magnifiers, speech recognition software, text readers, etc.)

Autonomous Agents: Autonomous agents are software or devices which act independently without direct human control (e.g., unmanned orbiting satellites).

Autonomy: Autonomy is the ability to think and act for oneself without external intervention. It usually implies the presence of rational thought and judgment.

Bioethics: Bioethics is an area of applied ethics concerned with ethical implications within medicine and medical research.

Biometrics: Biometrics refer to devices and techniques to detect individual specific physiological or behavioural traits in order to verify an individual's identity (e.g., fingerprints, voice recognition, retinal scanning, DNA profiling, etc.)

Categorical Imperative: The categorical imperative was a universal principle from Kant which posited that one should so act that the maxim of their will could always hold for all people all the time.

Cloning (Somatic cell nuclear transfer): Cloning refers to asexual reproduction where the nucleus of an ovum is replaced with the nucleus of a somatic cell of an adult resulting in ovum development without the involvement of sperm.

Cognitive Science: Cognitive science is an interdisciplinary field of research focusing on the study of intelligent activity in living organisms and machines.

Computer Ethics: Computer ethics is an area of applied ethics that deals with the social and ethical aspects of computer technology and its use.

Confidentiality: Confidentiality refers to an obligation to not disclose personal or proprietary

Convergence: This refers to a globalization process where information and communication technologies become more closely intertwined to allow more fluid and ubiquitous interactions.

Cyber Identity Theft: Cyber identity theft refers to the online appropriation of identity information without the permission of the individual in question.

Cyber Porn: Cyber porn describes Sexually explicit material (e.g., images, text, video) found on the Internet.

Cyber-Bullying: Cyber-bullying (or cyber-stalking) refers to online harassment in online postings or conversations intended to humiliate and intimidate the receiver.

Cyber-Crime: Cyber-crime refers to crimes conducted through the use of computers and networking technology (I.e. information theft, online banking fraud, etc)..

Cyber-Stalking: Cyberstalking refers to online activities intended to threaten or harm someone, often by accessing confidential information to use against someone.

Cyber-Victimization: Cyber-victimization refers to the unfair treatment of someone due to cyber-criminal activity.

Cyborg: A cyborg (also cybernetic organism) refers to the augmentation of the human mind and body through technological enhancements. The resulting cyborg is an entity that is part human and part machine.

Digital Divide: The digital divide refers to the technology and Internet access gap that exists between peoples and societies.

Discourse Ethics: Discourse ethics is an area of philosophical inquiry and communication study focused on rational communication and normative values underlying discourse.

Disembeddedness: This is an outcome of globalization where actions and relations are removed from local contexts and restructured across time and space.

Divergence: This refers to a globalization process that highlights cultural uniqueness around the globe.

Email: Electronic mail is the exchange of messages by telecommunication.

Ethics: Ethics is the study of moral conduct and notions of what is right, wrong, good, bad, and of what ought and out not be done.

Externally-Sponsored Research: Externally-sponsored research refers to research conducted or controlled by an outside sponsor (i.e., Multinational research project carried out in a developing country). Research protocols are often designed in one country while test subjects are recruited from another country.

Gene Doping: Gene doping refers to the non-therapeutic use genetic material to improve athletic performance.

Genetic Risk: Genetic risk is a probabilistic indication that an individual may be predisposed or to develop a specific disease based on genetic testing.

Global Communication: Global communication refers to the worldwide communication achieved through the use of information and communication technologies to cross national boundaries.

Global Consciousness: This is a globalization process globalization that entails an awareness of other cultures and social practices around the globe.

Global Organization: This describes an organization aligned with a global system and economy rather than with any one nation or local economy.

Globalization: This describes the developing processes of interconnections between societies, cultures, and organizations within a relational system of technology that defines contemporary society.

Human Cloning: Human cloning refers to the creation of identical human embryos and tissues in to harvest stem cells for research and transplantation purposes.

IGO. This refs to intergovernmental organizations (e.g., UN, NATO).

In Vitro Fertilization (IVF): IVF is an assisted reproductive procedure where female ova (eggs) are extracted and fertilized with a male's sperm in a Petri dish.

Informational Privacy: Refers to the security of identifiable patient health information and clinical data found in patient records and in communications among healthcare professionals and patients.

Informed Consent: Informed consent is a research protocol where subjects voluntarily confirm their willingness to participate in a particular research project. This is documented using a signed and dated informed consent form.

INGO: This refers to international nongovernmental organizations (e.g., Amnesty International, Red Cross)

Intellectual Property (IP): IP refers to the laws protecting the property/author rights (often for a limited time) over their creative works or inventions.

Knowledge Revolution: The knowledge revolution refers to the socioeconomic change from an industrial age focusing on product manufacturing to a knowledge age focused on creating and using knowledge.

Laissez-Faire Capitalization: This refers to the assumption that a free market economy has adequate checks and balances to ensure that the interests of all members of society will be met and that economic prosperity emerges from free markets with limited government intervention.

Law of Technoethics: The Law of Technoethics holds that ethical rights and responsibilities assigned to technology and its creators increases as technological innovations increase their social impact.

Malware: Malware are software designed to create negative within some information technology (E.,g., computer virus).

Media Ethics: Media ethics is an area of applied ethics concerned with ethical principles and media standards.

Medical Ethics: Medical ethics is an area of applied ethics concerned with moral values in medicine within situations where moral values are implicated.

Multiculturalism: This refers to how different cultures coexist within a society. It seeks to ensure that cultural diversity is protected and nurtured.

Multinational Organization: This refers to an organization aligned with one nation while doing business internationally.

Nanotechnology: Nanotechnology refers to the manipulation of matter at the level of atoms and molecules.

Netiquette (Internet Etiquette): Netiquette refers to Internet guidelines designed to ensure civility in online discussions and message postings.

Philosophy of Technology: The Philosophy of Technology is a branch of philosophy that deals with the intellectual study of technology and its social implications.

Phishing: Phishing is a deception using email to coerce victims to disclose personal information about themselves (E.g. credit card numbers, back account information, passwords, etc.) for fraudulent purposes.

Plagiarize: Plagiarism is stealing and passing off the ideas or words of another without crediting the source.

Privacy: Privacy is a fundamental human right of individuals to control access to parts of their personal life without interference from others.

Reprogenetics: Reprogenetics refers to a variety of new medical research applications emerging from the interconnection of reproductive medicine and genetic technologies (e.g., Embryonic stem cell research, genetic manipulation of embryos, embryo cloning, genetic diagnosis, etc.).

Research Protocol: Research protocols are guidelines describing all procedural and organizational aspects of the proposed research project including, research rationale, objectives, methodology, participant role, ethical research procedures, etc.

Software Piracy: Software Piracy is the unauthorized duplication and/or distribution of software in violation of intellectual property laws.

Spam: Spam refers to he unsolicited bulk distribution of promotional emails.

Telehealth: Telehealth describes the use of ICT to share and to maintain patient health information and to provide clinical care and health education to patients and professionals when distance separates the participants.

Time and Space Compression: This refers to a globalization process where information and communication technologies changes communication patterns and speeds up interactions.

UNESCO: UNESCO is an international agency within the United Nations which deals with developing and overseeing educational programs and project within developing areas of the world.

Veil of Ignorance: The veil of ignorance, is a notion advanced by John Rawls describing a an individual stance towards ethical decision making where individuals attempt to make choices without knowing their personal potential benefits/costs and social circumstances.

Virtual Alliance: A virtual alliance is a network of organizations or individuals linked by computer-mediated communication and collaboration.

Vulnerable Subjects: Vulnerable subjects are research subjects whose decision to participate or not is at risk of being unduly influenced by expectations of benefits or expectation of a retaliatory response.

Web 2.0: Web 2.0 refers to interactive Internet applications developed to provide users with increased access, control, information sharing,and collaboration (E.g.,. wikipedia, facebook myspace, etc.).

About the Authors

Rocci Luppicini, PhD is an Assistant Professor in the Department of Communication at the University of Ottawa in Canada. He is co author of the *Handbook of Research on Technoethics* (with Dr. R. Adell), the first comprehensive reference work in the English language. He has published in a number of other areas including virtual learning communities and practice (Quarterly Review Of Distance Education), research methodology on online instruction (Journal of Distance Education), issues in Higher Education, instructional design (Canadian Journal of Learning and Technology), and design research (International Journal of Technology and Design Education). Recent authored and edited books include, *Online Learning Communities in Education* (2007), the *Handbook of Conversation Design For Instructional Applications* (2008), and *Trends in Educational Technology and Distance Education in Canada* (2008).

Alison Anderson, PhD, is Professor in Sociology at the University of Plymouth, UK. Prior publications include Media, Culture and the Environment (University College London and Rutgers University Press, 1997). Her recent articles on journalistic portrayals of environmental risks, nanotechnologies, genetics and terrorism have appeared in Science Communication, Knowledge, Technology and Society, Genetics and Society and Health, Risk and Society.

Deb Gearhart, PhD, recently became the director of eCampus for Troy University. Previously Deb served as the founding Director of E-Education Services at Dakota State University in Madison, South Dakota and was there for the 11 years. Before joining Dakota State she spent 10 years with the Department of Distance Education at Penn State. Deb is an associate professor for educational technology at Dakota State University teaching at both the undergraduate and graduate levels.

She has co-authored at textbook entitled Designing and Developing Web-Based Instruction. Dr. Gearhart has earned a BA in Sociology from Indian University of Pennsylvania. She earned a M.Ed. in Adult Education with a distance education emphasis and an M.P.A. in Public Administration, both from Penn State. Deb completed her Ph.D. program in Education , with a certificate in distance education, from Capella University.

A. Pablo Iannone, PhD, is Professor of Philosophy at Central Connecticut State University, studied engineering, mathematics, philosophy and literature at the Universidad Nacional de Buenos Aires, received a B. A. in philosophy from U.C.L.A. and an M.A. and a Ph.D. in philosophy from the University of Wisconsin-Madison, and pursued graduate studies in business and economics at the University of Wisconsin-Madison and Iowa State. He taught at Canada's Dalhousie University, Perú's Universidad Inca Garcilaso de la Vega, and the U.S. universities of Wisconsin-Madison, Texas at Dallas, Iowa State, and Florida,. His publications include nine philosophy books, two literature books, and articles and reviews.

Mathias Klang, PhD, currently holds positions both at the University of Lund and the University of Göteborg. In Lund he is conducting a copyright research project aimed at developing the state of Open Access at university libraries in Sweden. In Göteborg Mathias conducts research in the field of legal informatics with particular interest in copyright, democracy, human rights, free expression, censorship, open access and ethics. He has published several articles in these topics.

Lorenzo Magnani, PhD, is a Professor in the Department of Philosophy and Computational Laboratory at the University of Pavia in Italy and the Department of Philosophy at Sun Yat-sen University in China.

Lynne Roberts, PhD, is a Research Fellow at the Crime Research Centre at the University of Western Australia. Lynne has conducted both quantitative and qualitative research in and about virtual communities. Based on this research she has had 13 journal articles and book chapters published on social interaction, relationship formation, sense of community and gender-switching online; and ethical and methodological issues in conducting research in virtual environments.

Neil C. Rowe, PhD, is Professor and Coordinator of Research in Computer Science at the U.S. Naval Postgraduate School where he has been since 1983. He has a Ph.D. in Computer Science from Stanford University (1983), and E.E. (1978), S.M. (1978), and S.B. (1975) degrees from the Massachusetts Institute of Technology. His main research interest is the role of deception in information processing, and he has also done research on intelligent access to multimedia databases, image processing, robotic path planning, and intelligent tutoring systems. He is the author of over 140 technical papers and a book.

Martin Ryder, M.A, is a development engineer for Sun Microsystems. He also serves on the adjunct faculty at the University of Colorado at Denver, teaching research methods in Information and Learning Technologies.

May Thorseth, PhD, Associate Professor at the Department of Philosophy, NTNU Norwegian University of Science and Technology. Director of Programme for Applied Ethics, NTNU. Main research field is political philosophy, ethics and argumentation theory in the context of globalisation and multicultural conflicts. Ongoing research project: "Worldwide Deliberation and Public Use of Reason Online". Current member of the Regional Research Ethical Committee in Medicine in Mid-Norway, the National Research Ethical Committee in Medicine in Norway, and previous member of the Clinical Ethics Committee at St. Olavs Hospital, Trondheim.

Index